P9-CZV-584

The Ten Most Beautiful Experiments

ALSO BY GEORGE JOHNSON

Miss Leavitt's Stars: The Untold Story of the Woman Who Discovered How to Measure the Universe

A Shortcut Through Time: The Path to the Quantum Computer

Strange Beauty: Murray Gell-Mann and the Revolution in Twentieth-Century Physics

Fire in the Mind: Science, Faith, and the Search for Order

In the Palaces of Memory: How We Build the Worlds Inside Our Heads

Machinery of the Mind: Inside the New Science of Artificial Intelligence

Architects of Fear: Conspiracy Theories and Paranoia in American Politics

THE TEN MOST BEAUTIFUL EXPERIMENTS

George Johnson

ALFRED A. KNOPF NEW YORK 2008

THIS IS A BORZOI BOOK
PUBLISHED BY ALFRED A. KNOPF

www.aaknopf.com

Library of Congress Cataloging-in-Publication Data

Johnson, George, [date]
The ten most beautiful experiments / by George Johnson.—1st ed.
p. cm.
ISBN 978-1-4000-4101-5
Includes bibliographical references and index.
1. Science—Experiments. I. Title.
Q182.3.J65 2008
507.8—dc22 2007027839

Manufactured in the United States of America
First Edition

When Albert Einstein was an old man and sat down to write a short volume of autobiographical notes—"something like my own obituary"—he remembered the day his father showed him a compass. Turning it this way and that, the boy watched in wonder as the needle pointed insistently north. "I can still remember—or at least believe I can remember—that this experience made a deep and lasting impression upon me," Einstein wrote. "Something deeply hidden had to be behind things."

CONTENTS

PROLOGUE

ON A CLEAR winter morning several years ago, I drove up the hill to St. John's College to play with electrons. I'd recently met the president of the school, which is nestled in the splendid isolation of the Santa Fe foothills, and was impressed to learn that the students, as part of their studies in the humanities, were expected to reenact the famous experiment of 1909 in which Robert Millikan isolated and measured these fundamental particles, showing them to be bits of electricity.

St. John's, like its sister college in Annapolis, pursues a classical curriculum, with physics starting around 600 BC with the Presocratic philosophers. That was when Thales of Miletus made the first stab at a Grand Unified Theory: "Everything is made of water." Today he would probably be working on superstrings.

Thales had also noticed that a rock called magnetite, found in the province of Magnesia, exerted an invisible pull on metal and that rubbing a piece of amber, a substance the Greeks called *elektron,* gave it a mysterious charge: it attracted pieces of straw and chaff. More than two thousand years later William Gilbert, Queen Elizabeth I's physician, noted that glass rubbed with silk became "amberized"—*electrified* (he was the first to use the term)—and that other materials could also be enlivened this way. Friction, Gilbert speculated, heated some kind of watery humor giving rise to a sticky, vaporous "effluvium" of charge. A French chemist, Charles-François de Cisternay Dufay, went on to discover that rubbed amber repelled objects that rubbed glass attracted. Electricity, he concluded, must come in two forms: "resinous" and "vitreous." Something deeply hidden lay behind things. Millikan found a way to get a grip.

I FOUND the physics lab in the basement of a two-story Territorial-style building fronted by a long white veranda and surrounded by pines. Class was not in session, and the shades had been pulled, the lights turned low. At the far side of the room, the laboratory director, Hans von Briesen, was assembling electronic components on a wooden laboratory table. One of the customs at St. John's is that students and teachers (tutors, they are called) address one another with honorifics—Mr. von Briesen, Mr. Johnson—making hallway conversations sound a little like the *New York Times*.

The idea of Millikan's experiment, Mr. von Briesen explained, was to use a perfume atomizer to spray minuscule droplets of oil into a space between two metal plates, one charged resinously and the other vitreously. Rubbed by the air, some of the drops, like Thales's amber, would become electrified. By varying the voltage across the two plates, a droplet could be made to move up and down, or with just the right touch to hover suspended in midair.

From the mass of the droplet and the amount of voltage required to counteract its fall, you can determine its charge. Measure enough droplets and you can see whether charge, like a fluid, comes in any amount whatsoever or, like pocket change, only in discrete quantities. If the latter is true then the smallest amount would be the elementary unit of electricity—the charge of the electron.

When the setup was complete and the room darkened the experiment began. After several trial runs Mr. von Briesen invited me to take a look. Gazing into the chamber through a magnifying eyepiece—a little telescope—I saw the droplets. Illuminated from behind, they shone like a constellation or galaxy. Millikan himself had described them this way: "The appearance of this drop is that of a brilliant star."

SCIENCE in the twenty-first century has become industrialized. The experiments so often celebrated in the newspapers—sequencing the genome, proving the existence of the top quark, discovering a new planet by analyzing the wobble of a distant star—cost millions of dollars. They generate terabytes of data to be analyzed by supercomputers: calculating factories spewing so much heat that they are equipped with cooling stacks that consume the energy of small towns. The experiments are carried out by research teams that have grown to the size of corporations.

But until very recently the most earthshaking science came from individual pairs of hands. From a single mind confronting the unknown. The great experiments that mark the edges of our understanding were most often performed by one or two scientists and usually on a tabletop. Computation, if there was any, was carried out on paper or later with a slide rule.

These experiments were designed and conducted with such straightforward elegance that they deserve to be called beautiful. This is beauty in the classical sense—the logical simplicity of the apparatus, like the logical simplicity of the analysis, seems as pure and inevitable as the lines of a Greek statue. Confusion and ambiguity are momentarily swept aside and something new about nature leaps into view.

As a science writer, I have most often been attracted to airy edifices like quantum mechanics or general relativity, which seek to capture reality with a few courtly laws. For a sign of just how abstract this quest has become, one need look no further than superstring theory, which posits that matter is ultimately generated by mathematical snippets vibrating in ten-dimensional space. This is fascinating stuff, but so rarefied and confusing—so far over my, or maybe anyone's, head—that I began to feel a need for basics.

The magazine *Physics World* once conducted a survey asking its readers what they considered the most beautiful of all experiments. From the results, a roster was compiled of the top ten, all predictably within the realm of physics. But what, I wondered, if one were to cast the net wider? I decided to make my own list.

The question was where to begin. With Thales rubbing amber to create static electricity? That lacked the kind of elegance I was looking for. There were no controls, no systematic attempt to see what materials, under what conditions, could be charged this way. As Gilbert went on to show, there was nothing unique about amber. With Thales experimental science had not yet begun.

How about Pythagoras, another of the Presocratics, who discovered that the musical notes sounded by a plucked string correspond to precise mathematical ratios? If the whole string sounds a perfect C, three-fourths of the string will sound an F and two-thirds a G. Pinch the string in half and it will sound a C again, an octave higher. All is number,

Pythagoras declared—another Grand Unified Theory. He should have stopped while he was ahead. Fire, he went on to speculate, is made of twenty-four right-angle triangles, surrounded by four equilaterals, which are made in turn of six triangles. Air is composed of forty-eight triangles, water of one hundred and twenty. Experiment gave way to mysticism.

Another candidate might have been Archimedes. The dubious legend about his jumping from a bathtub shouting "Eureka," having discovered the physical law of buoyancy, trivializes the grandeur of his accomplishment. His treatise *On Floating Bodies* is considered a masterpiece of mathematical reasoning, and not just because of its derivation of Archimedes's principle (a body submerged in a fluid is acted upon by an upward force equal in magnitude to the weight of the fluid displaced). He also figured out, from first principles, how a cone-shaped object called a paraboloid would float if immersed in water. (Icebergs are roughly paraboloid and behave pretty much as Archimedes said.)

His greatness, however, lay more in reasoning than in experiment. Another great theorist. What I was looking for were those rare moments when, using the materials at hand, a curious soul figured out a way to pose a question to the universe and persisted until it replied. Ideally the apparatus itself would be a thing of beauty, with polished wood, brass, shining black ebonite. More important would be the beauty of the design and the execution, the cleanness of the lines of thought.

For that I had to jump from ancient Greece all the way to the seventeenth century, when a man named Galileo coaxed out a fundamental law of motion. From there, I proceeded step by step, visiting nine more stops on the scientific trail, eventually meeting up again with Millikan and his tiny stars.

Likelier than not, anyone who reads this book could come up with a different list. "Shouldn't you just call it *Ten Beautiful Experiments*?" a friend objected. Probably so. But I hope that there is art in the arbitrariness, both in my selection of the experiments and in what I have chosen to tell about each one. This is not a book about great discoveries, the serendipitous surprises like Galileo's spying of satellites circling Jupiter or Charles Darwin's observations about finches. Those were not the kind of deliberate, controlled interrogations of reality that I wanted to explore. Nor is this intended as a collection of miniature scientific biographies—there are already plenty of good ones. Some lives, like those of Antoine-Laurent Lavoisier and Albert Michelson, diverted me with their strange details. Others, like Galileo's and Newton's, have been told too many times before. I've tried to sketch each scientist with a charcoal wash. I want the experiment, not the experimenter, to be the protagonist.

To keep the stories as crisp as possible, I've spent little ink trying to parcel out credits, fighting the historians' fights. James Joule's surprising discovery about energy and heat was anticipated by Robert Mayer, but it was Joule who did the beautiful experiment. I like what Lord Kelvin had to say about that: "Questions of personal priority, however interesting they may be to the persons concerned, sink into insignificance in the prospect of any gain of deeper insight into the secrets of nature."

The Ten Most Beautiful Experiments

CHAPTER 1

Galileo

The Way Things Really Move

Galileo Galilei, by Ottavio Leoni

> It is very unpleasant and annoying to see men, who claim to be peers of anyone in a certain field of study, take for granted certain conclusions which later are quickly and easily shown by another to be false.
>
> —Salviati, in Galileo, *Two New Sciences*

WHEN you throw a rock, catch a ball, or jump just hard enough to clear a hurdle, the older, unconscious part of the brain, the cerebellum, reveals an effortless grasp of the fundamental laws of motion. Force equals mass times acceleration. Every action results in an equal and opposite reaction. But this ingrained physics is sealed off from the newer, upper brain—the cerebrum, seat of intelligence and self-awareness. One can leap as gracefully as a cat but be just as powerless to explain the inverse square law of gravity.

Aristotle, in the fourth century BC, made the first ambitious attempt to articulate the rules of motion. An object falls in proportion to its weight—the heavier a rock, the sooner it will reach the ground. For other kinds of movement (pushing a book across a table or a plow across a field), a force must be constantly applied. The harder you push, the faster the object will go. Stop pushing and it will come to a halt.

It all sounds eminently sensible and obvious and, of course, is exactly wrong.

What if you place the book on a sheet of ice and give it a gentle shove? It will keep moving long after the impetus is removed. (Asked why an arrow keeps going after it leaves the bowstring, the Aristotelians said that it was pushed along by the incoming rush of air.) Now we know that something set in motion stays in motion until stopped by something else, or worn down by friction. And a one-pound weight and a five-pound weight, dropped at the same moment, will fall side by side to the ground. Galileo showed it was so.

It's entirely predictable that the great debunker of Aristotle—celebrated in a play by Bertolt Brecht, an opera by Philip Glass, and a pop song by the Indigo Girls—would come in for his own debunking. It is doubtful, historians tell us, that Galileo dropped two weights from the Leaning Tower of Pisa. Nor do they believe that he hit on his insight about pendulums—that each swing is of equal duration—while watching a certain chandelier in the cathedral of Pisa and timing it with his heartbeat.

His credentials as a cosmologist have also dimmed under

scrutiny. Galileo was the most eloquent advocate of Copernicus's sun-centered solar system—his *Dialogue Concerning the Two Chief World Systems* is the first great piece of popular science writing—but he never accepted Kepler's crucial insight: that the planets move in ellipses. The orbits, Galileo assumed, had to be perfect circles. Here he was following Aristotle, who proclaimed that while motion on Earth (in the "sublunar" realm) must have a beginning and an end, celestial motion is necessarily circular.

For that to be true and match what was happening in the sky, the planets would have to move not just in circles but in circles within circles—the same old epicycles that had weighed down Ptolemy's geocentric universe. Galileo brushed off the problem. Most disappointing of all, he probably did not, as legend has it, follow his forced apology to the Inquisitors of Rome by muttering under his breath, *Eppur si muove,* "And yet it moves." He was no martyr. Knowing he had been beaten, he retired to the solitude of Arcetri to lick his wounds.

Galileo's strongest claim to greatness lies in work he did long before his troubles with the Vatican. He was studying nothing so grand as stars or planets but the movement of simple, mundane objects—a subject far more perplexing than anyone had imagined.

Whether or not the research actually began at the Tower of Pisa hardly matters. He described a similar experiment in his other masterpiece, *Discourses Concerning Two New Sciences,* completed during his final years of exile. Like the earlier work it is cast as a long conversation among three Italian noblemen—Salviati, Sagredo, and Simplicio—who are trying to understand how the world works.

Salviati is the stand-in for Galileo, and on the first day of the gathering he insists that, dropped simultaneously, a cannonball weighing 100 pounds and a musket ball weighing 1 pound will hit the ground at almost the same time. In an experiment, he concedes, the heavier one did in fact land "two finger-breadths" sooner, but Salviati recognized that other factors, like air resistance, muddied the results. The important point was that the impacts were *almost* in unison: when the cannonball hit the ground, the musket ball had not traveled just 1/100 the distance—a single cubit—as common sense would have predicted. "Now you would not hide behind these two fingers the ninety-nine cubits of Aristotle," he chided, "nor would you mention my small error and at the same time pass over in silence his very large one." All other things being equal, the speed at which an object falls is independent of its weight.

A harder question was what happened between the time a ball was released and the time it struck the ground. It would pick up speed along the way—everybody knew that. But how? Was there a large spurt of motion at the beginning, or a lot of little spurts continuing all the way down?

With nothing like time-lapse photography or electronic sensors to clock a falling body, all you could do was speculate. What Galileo needed was an equivalent experiment, one in which the fall would be slower and easier to observe: a ball rolling down a smooth, gentle plane. What was true for its motion should be true for a steeper incline—and for the steepest: straight down. He had found a way to ask the question.

The year was probably 1604. Three decades later he, or rather Salviati, described the thrust of the experiment:

> A piece of wooden moulding or scantling, about 12 cubits long, half a cubit wide, and three finger-breadths

> thick, was taken. On its edge was cut a channel a little more than one finger in breadth. Having made this groove very straight, smooth, and polished, and having lined it with parchment, also as smooth and polished as possible, we rolled along it a hard, smooth, and very round bronze ball.

A scantling is a piece of wood, and a Florentine cubit was twenty inches, so we can imagine Galileo with a twenty-foot-long board, ten inches wide, propping it up at an angle.

> Having placed this board in a sloping position, by lifting one end some one or two cubits above the other, we rolled the ball, as I was just saying, along the channel, noting, in a manner presently to be described, the time

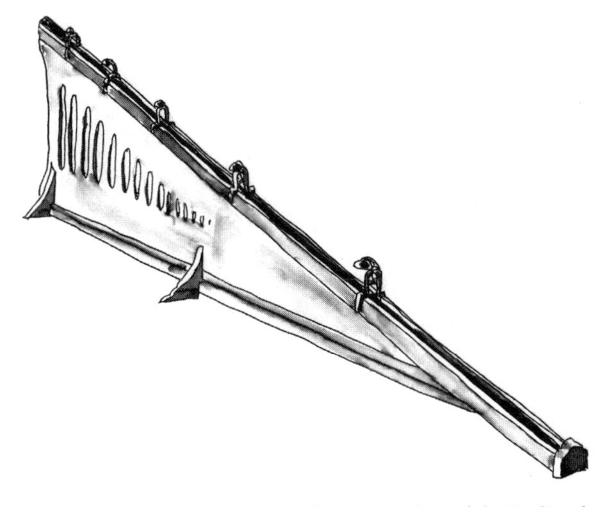

An early-nineteenth-century demonstration of the inclined plane experiment. The rolling ball causes the bells to ring.
Drawing by Alison Kent

> required to make the descent. We repeated this experiment more than once in order to measure the time with an accuracy such that the deviation between two observations never exceeded one-tenth of a pulse-beat.

Once they had perfected the technique, Salviati went on to explain, they timed how long it took the ball to traverse one-fourth of the track, then two-thirds, then three-fourths. They repeated the experiment with the board set at different slopes—100 measurements in all. These were taken with a simple device called a water clock, essentially an hourglass that parcels out seconds with liquid instead of sand:

> We employed a large vessel of water placed in an elevated position. To the bottom of this vessel was soldered a pipe of small diameter giving a thin jet of water, which we collected in a small glass during the time of each descent, whether for the whole length of the channel or for a part of its length. The water thus collected was weighed, after each descent, on a very accurate balance. The differences and ratios of these weights gave us the differences and ratios of the times, and this with such accuracy that although the operation was repeated many, many times, there was no appreciable discrepancy in the results.

The weight of the water was equivalent to the passage of time. Ingenious. But maybe, some modern historians have concluded, a little too good to be true. Reading Galileo's words some three centuries later, Alexandre Koyré, a professor at the Sorbonne, could barely contain his scorn:

> A bronze ball rolling in a "smooth and polished" wooden groove! A vessel of water with a small hole through which it runs out and which one collects in a small glass in order to weigh it afterwards and thus measure the times of descent . . . What an accumulation of sources of error and inexactitude! It is obvious that the Galilean experiments are completely worthless.

Koyré suspected that there had been no experiment—that Galileo was using an imaginary demonstration with rolling balls as a pedagogical device, an illustration of a law of physics that he had figured out mathematically, through pure deduction, the old-fashioned way. Galileo, it seemed, had been debunked again.

Koyré was writing in 1953. Twenty years later Stillman Drake, one of the leading experts on Galilean science, was sleuthing among the manuscripts in the Biblioteca Nazionale Centrale in Florence when he came across some unpublished pages—entries from Galileo's own notebook.

Galileo was something of a pack rat, and when his notebooks were published around the turn of the twentieth century, the editor, Antonio Favaro, had left out some pages that appeared to be no more than scribbles, a mess of calculations and diagrams that didn't make sense. The pages were apparently out of order, with little clue as to when they had been written or what their author was working on.

Drake was researching a new English translation of *Two New Sciences.* For three months at the beginning of 1972, he sat in Florence going through 160 pages of the seventy-second volume of Galileo's papers, comparing watermarks and styles of handwriting, restoring the pages to what

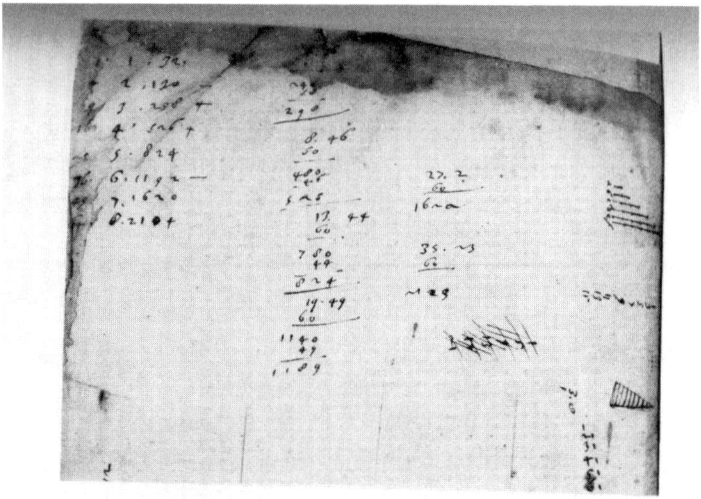

A page from Galileo's notebook

seemed a sensible order. Among the earliest were what appeared to be data from the experiment of 1604, when Galileo was in Padua.

From the jottings, Drake re-created the centuries-old experiment, and with just a little license, we can imagine what was going through Galileo's mind. He releases the ball at the top of the wooden incline noting that in the first few moments, it travels a distance of 33 *punti*, or points. (Galileo was using a ruler marked into sixty equal units, and a point, Drake deduced, was just shy of one millimeter.) After an equal amount of time has passed, the ball, picking up speed, covers a total of 130 *punti*, and by the end of the third interval, 298 *punti*. Then 526, 824, 1,192, 1,620 . . . faster and faster. These were real data. For the final distance, when the ball would have been moving at top speed, Galileo had originally written 2,123 *punti*, scratching it out and correcting it to

2,104. By some of his figures, he put a plus or a minus sign, apparently indicating when his measurements seemed high or low.

The units of time he was using don't matter. We might as well call them ticks. The important thing is that each interval be the same:

1	2	3	4	5	6	7	8	ticks *(time)*
33	130	298	526	824	1,192	1,620	2,104	*punti (accumulated distance)*

At first no pattern leaps forth. With each tick the ball covers more ground, but by what rule? Galileo started playing with the numbers. Maybe the speed increased according to some arithmetical progression. What about alternating odd numbers: 1, 5, 9, 13, 17, 21 . . . ? On the second tick the ball would move five times faster than on the first tick, covering 5 × 33 or 165 *punti.* Too high but maybe within the range of experimental error. The distance covered on tick three would be nine times greater: 33 × 9 = 297 *punti.* Right on the mark! And on the fourth tick 13 × 33 = 429. Too low. Then 17 × 33 = 561, too high. And 21 × 33 = 693, way too low. . . . Drake could see on the manuscript page where Galileo scratched out the numbers to try again.

On the first tick the ball had covered 33 *punti,* then 130. What if you divide the numbers? 130/33 = 3.9. The distance had increased almost four times. With the third tick, the increase was 298/33, slightly more than nine times the initial distance. Then 15.9, 25.0, 36.1, 49.1, 63.8. He rounded the numbers and wrote them, using a different ink and pen, in a column: 4, 9, 16, 25, 36, 49, 64.

He had found the key: allowing for a bit of error, the distance covered increased with the *square* of the time. With a longer board, one could confidently predict that on the next tick the factor would be 81 (9^2) and then 100, 121, 144, 169. . . . That Galileo's numbers were not exact testified to the reality of the experiment. That they were as close as they were testified to his skill as an experimenter.

In these calculations the distances are cumulative: by the fourth tick the ball has traversed a total of sixteen times the distance it covered at the end of the first tick. But how far does it travel during each separate interval, between ticks three and four compared with ticks two and three? The answer can be backed out with arithmetic.

It is the nature of squares that they are the sums of the odd numbers that precede them: $4 = 1 + 3$; $9 = 1 + 3 + 5$; $16 = 1 + 3 + 5 + 7$. Implicit in the times-square law is that the distances between ticks must increase according to the progression of odd numbers. Galileo's data show how this works.

1	2	3	4	5 . . .	ticks *(time)*
33	130	298	526	824 . . .	*punti (accumulated distance)*
	130–33	298–130	526–298	824–526	
	97	168	228	298	*punti (distance traveled in an interval)*
	97/33	168/33	228/33	298/33	
	2.9	5.1	6.9	9.0	*ratio of distances*

Tick by tick the ball travels three times the distance, then five times, then seven, then nine. In fact Galileo could have started with the odd-number progression and derived the times-squared relationship. However he did it, the result was a fundamental new law. The steeper the slope, the faster the ball would roll, but always according to the same rule—which would presumably hold if the slope was ninety degrees, straight down.

At the other extreme, a slope of zero degrees, there would be no acceleration. Once the ball, traveling down the incline, reached the flat tabletop, it would begin moving at a uniform speed—forever if the plane was infinite and friction didn't interfere. And if the moving ball reached the edge of the table and dropped off? On the triumphant fourth day of *Two New Sciences,* Galileo provides the answer: the unhurried horizontal motion and the downwardly accelerated vertical motion combine to yield the familiar parabolic shape of a projectile.

There was still the question of how Galileo did such precise timing, working with intervals of less than a second. Using a flowerpot as a water clock, a Cornell University graduate student, Thomas B. Settle, rolled billiard balls down a two-by-six pine plank and, once he had tuned his reflexes, demonstrated the validity of the times-squared law. But both he and Drake doubted that someone starting from ignorance could have discovered the relationship with so crude an apparatus. Galileo's technique, Drake proposed, was more brilliant and surprising.

It wouldn't have been necessary, he realized, for Galileo to clock time the modern way—in seconds, half seconds, or any other conventional measure. All that was needed was a way to *divide* time into equal portions, and this, Drake recognized, is a talent that comes naturally to any good musician.

"The conductor of an orchestra, moving his baton, divides time evenly with great precision over long periods without thinking of seconds or any other standard unit," Drake wrote. "He maintains a certain even beat according to an internal rhythm, and he can divide that beat in half again and again with an accuracy rivaling that of any mechanical instrument." The same goes for the musicians and even for the audience. "If the cymbalist in the orchestra were to miss his entry by a tiny fraction of a second, say by a 64th note in the music, everyone would notice it, not just the conductor."

So, Drake speculates, this is what Galileo did: before the ball rolled down the incline, he established a rhythm by singing a simple tune. Drake tried the experiment with "Onward Christian Soldiers," at about two beats per second. Releasing the ball at the top of the incline, he used chalk to mark its position at each upbeat.

> ONward CHRIStian SO-ol-DIER-rs MARCHing AS to . . .

Like Drake, Galileo probably hadn't caught them all on the first run, but after several attempts he would have marked off the track in approximately half-second intervals, noting with some satisfaction that the spacing became progressively greater—that the ball, in a lawful manner, rolled faster and faster down the hill.

The next step was to tie a piece of catgut at each chalk mark, like the movable frets on the neck of a lute, an instrument Galileo knew how to play. Drake used rubber bands. Rolling the ball again and again, he listened as it struck the frets, adjusting their placement until the rhythm of the clicking was as uniform as a metronome's and in time with the march. When he was done, the frets showed precisely how far the ball had traveled during equal intervals of time. All that was left was to measure the spacing with a ruler.

Once Galileo had established his law, Drake believed, he showed it to others in an easier, less precise manner: by marking the track beforehand—1, 4, 9, 16, 25, 49, 64—and then using a water clock to confirm the timing. But that was a demonstration, not an experiment.

Why didn't he write about his original method? The best Drake could suggest is that Galileo was afraid of sounding silly. "Even in his day, it would have been foolish to write, 'I tested this law by singing a song while a ball was rolling down a plane, and it proved quite exact.' " It wasn't long before he had picked up his telescope and moved on to other things.

Today, more than three hundred years after his death, visitors to the Museo di Storia della Scienza, the history of science museum in Florence, can see one of the withered fingers that picked up the metal ball each time it reached the bottom of the incline, returning it to the top for another ride. It was

Galileo's finger

removed by an admirer, along with a tooth, the fifth lumbar vertebra, and a couple of other fingers, when Galileo's body was exhumed, a century after his death, to be moved to a better burial site. Preserved in a reliquary like the bone of a saint, the long, thin finger has been mounted so that it points upward, as though beckoning to the sky.

CHAPTER 2

William Harvey

Mysteries of the Heart

William Harvey, by Willem van Bemmel

But what remains to be said upon the quantity and source of the blood which thus passes is of a character so novel and unheard-of that I not only fear injury to myself from the envy of a few, but I tremble lest I have mankind at large for my enemies, so much doth wont and custom become a second nature. Doctrine once sown strikes deep its root, and respect for antiquity influences all men. Still the die is cast, and my trust is in my love of truth and the candour of cultivated minds.

—*William Harvey*

THE CHICK embryo lying in a container of tepid water looked like a little cloud. Its shell had been carefully peeled away, and inside there throbbed a minuscule heart—a red dot no bigger than a pinpoint that disappeared and reappeared with every beat. Years later, in 1628, a London physician named William Harvey described the phenomenon: "Betwixt the visible and invisible, betwixt being and not being, as it were, it gave by its pulses a kind of representation of the commencement of life."

Probably no one had ever studied so many different kinds of hearts—dog hearts, pig hearts, the hearts of frogs, toads, snakes, fishes, snails, and crabs. A certain kind of shrimp found in the ocean and in the river Thames had a transparent body, and Harvey and his friends would watch its heart gyrate "as though it had been seen through a window." Sometimes he would remove a creature's heart altogether, feeling the slowing rhythm as it beat its last beats in his hand.

Observation by observation, Harvey was persuading himself—and hardly anybody else—that the great Galen, physician to gladiators and Roman emperors, was wrong. Galen had written, in the second century AD, that there were two kinds of blood carried by what amounted to two different vascular systems. A vegetative fluid, the elixir of nourishment and growth, was made in the liver and coursed through the body's web of bluish-colored veins. At the same time, a bright red vital fluid traveled through another network—the heart and arteries—activating the muscles and stimulating motion. (In the brain some of this vital fluid was turned into an ethereal essence that flowed through the nerves.) All the fluids were imbued with invisible pneuma, spirits that entered through the lungs with each breath before passing into the heart through a thick tube called the pulmonary vein. One thousand four hundred years later, this is what students were still being taught in medical school.

Harvey's indoctrination had probably begun at Cambridge, where in 1593 he entered Gonville and Caius College at the age of sixteen. The school's namesake, Dr. John Caius, a committed Galenist, had arranged for a royal charter granting the school two executed criminals each year for dissection and study. Along with his lessons on rhetoric, classics, and philosophy, Harvey had glimpses of human anatomy. The subject must have piqued his interest. From Cambridge he went on to the University of Padua, the most prestigious medical school in Europe.

Protected by the republic of Venice, the university felt freer than most to challenge Vatican dogma. At the time of Harvey's arrival, Galileo was teaching there, as was Hieronymus Fabricius, the greatest of Europe's anatomists. Each October on Saint Luke's Day (the corpses lasted longer in the cooler weather), the medical lectures began with a high mass, after which students would perch in the tiered balconies of the anatomy theater to watch as Fabricius and his assistants, scalpels in hand, gave a grand tour of the human interior.

After receiving his doctor's degree in 1602, Harvey returned to London, where he married the daughter of Lancelot Browne, the royal physician. Appointed to a position at Saint Bartholomew's, the city's oldest hospital, he established a practice whose patients would include Sir Francis Bacon, King James I, and James's successor, Charles I.

Though Harvey was short in stature and physically unimposing, his intense, dark eyes and raven hair must have made a formidable impression. The English writer John Aubrey described him as contemplative but choleric ("He was wont to say that man was but a great mischievous Baboon") and in the habit of wearing a dagger. That was the fashion, Aubrey acknowledged. "But this Dr. would be to apt to draw-out his dagger upon every slight occasion."

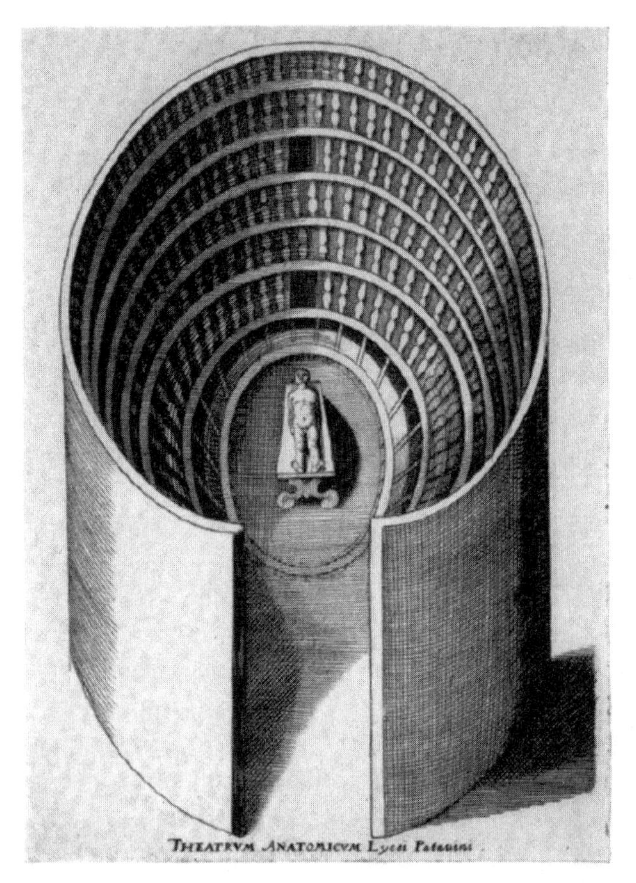

THE ANATOMY THEATRE OF FABRICIUS
Semidiagrammatic view, after Tomasini

Harvey's mind was like a scalpel. Whether he was making his rounds at the hospital or lecturing over a cadaver at the College of Physicians, no detail of human anatomy was small enough to escape his eye. When an organ differed from what Galenic wisdom prescribed, Harvey would diplomatically suggest that bodies must have changed since Galen's day. Privately he was piecing together a very different story.

He started with simple creatures, finding to his dismay that their hearts fluttered so rapidly that he could barely make sense of the motions. He knew there were two different kinds of beats: the systole, when the heart contracted, and the diastole, when it expanded. But when he viewed the process in vivo, it seemed impossible to distinguish one from the other.

> For I could neither rightly perceive at first when the systole and when the diastole took place, nor when and where dilatation and contraction occurred, by reason of the rapidity of the motion, which in many animals is accomplished in the twinkling of an eye, coming and going like a flash of lightning, so that the systole presented itself to me now from this point, now from that, the diastole the same. And then everything was reversed, the motions occurring, as it seemed, variously and confusedly together. My mind was therefore greatly unsettled, nor did I know what I should myself conclude, nor what believe from others. I was not surprised that Andreas Laurentius should have written that the motion of the heart was as perplexing as the flux and reflux of Euripus had appeared to Aristotle.

Laurentius was a Renaissance physician, and the Euripus was a strait along the Aegean coast of Greece where the tides move in and out seven times a day. Legend had it that Aristotle, dejected by his failure to understand these rhythms, drowned himself there.

If Harvey was to do better with the tides of the heart, he would need to observe the phenomenon at a slower pace, as Galileo did with his accelerating balls. In the "colder animals"—the amphibians, fish, reptiles, crustaceans, and mollusks—the heart beat more leisurely. These simple hearts presumably worked according to the same principles as those of mammals and men. In one experiment after another Harvey tuned his intuitions for the more difficult cases to come. For there were circumstances, he was soon to learn, in which even the metabolism of a warm-blooded animal slows to a crawl: during the final minutes of life when, weakened by vivisection, the poor creature's heartbeats become sparser and sparser until finally it gives up the ghost—or pneuma, or whatever had been keeping it alive.

THOUGH different in purpose and function, the two circulatory systems of Galen came within millimeters of each other in the heart. Carried by the superior and inferior vena cava, the bluish blood—constantly generated by the liver—flowed into and out of the heart's right-hand chambers. On the left side, sealed off by a thick wall called the septum, the red arterial blood flowed. Vessels also led to the lungs, which served to cool the blood and to carry pneuma—air—into the heart. It was there that the pneuma vitalized the venous blood, a tiny amount of which seeped across the septum through invisible pores and into the arterial ductwork.

Some of this picture had already been called into question. The Flemish physician Vesalius, in *Concerning the Fabric of the Human Body,* first published in 1543 (the same year as Copernicus's theory of heliocentrism), denied that blood could trickle across the heart's dividing wall. As hard as he looked he couldn't find even the tiniest pores. He was right for the wrong reason. We know now that bodily tissues are riddled with microscopic openings. It was Harvey who put the matter to rest: carefully cutting open an ox's heart, he poured water into the right side and noted that none made its way to the left.

Galen's followers also taught that the two kinds of blood—venous and arterial—moved like the tides, back and forth through the two systems. The vessels, animated by the vital spirit, expanded all at once, sucking up blood. When they contracted, the blood flowed the other way. The heart just went along for the ride, expanding and contracting like a bellows.

But that is not what Harvey was observing. When the heart contracted, on the systolic beat, like a hand bunching up into a fist, it became paler, as though blood was being squeezed out. When it expanded, on the diastole, it grew red again, as blood flowed back in. Even more telling, when he put his finger on an artery, he could feel it expand at the same time the heart contracted. The heart, it seemed, was driving the system. Galen had it backward. The push of contraction, not the pull of expansion, moved the blood. Cut an artery on a living mammal and blood came spurting out, "abundantly, impetuously, and as if it were propelled by a syringe."

If the heart was a pump, Harvey reasoned, he should be able to learn how it worked. Anatomists already knew that it was divided into four chambers. On top were the left and right auricles, below them the left and right ventricles. One day during a dissection Harvey placed a finger on a left ventricle. It expanded, filling with blood, just as the auricle above it contracted. Then, an instant later, the ventricle itself contracted, pushing blood out of the chamber and into the arteries. The same motions occurred on the right side. Again Galen was wrong. Blood was pumped not from right to left but from top to bottom: "These two motions, one of the ventricles, the other of the auricles, take place consecutively," Harvey wrote, "but in such a manner that there is a kind of harmony or rhythm preserved between them, the two concurring in such wise that but one motion is apparent."

He compared the movement to a machine's: "One wheel

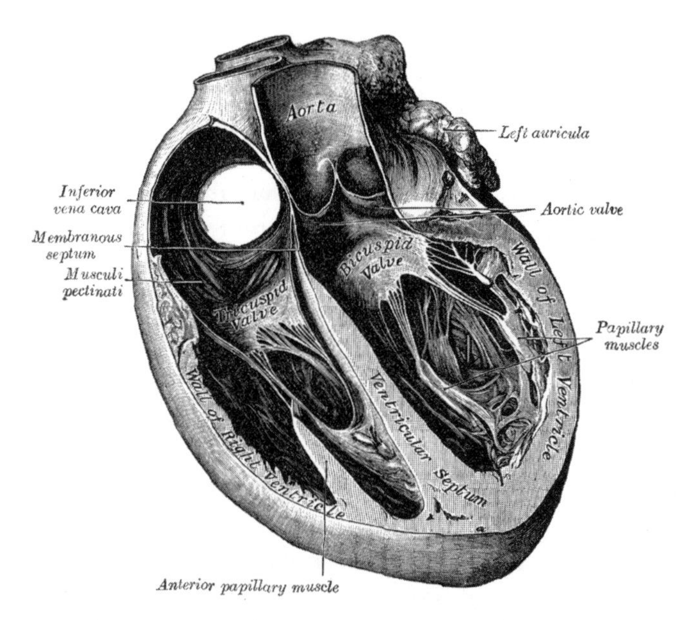

Cross section of a human heart from *Gray's Anatomy*

gives motion to another, yet all the wheels seem to move simultaneously." He knew that some of his readers might be offended by this mechanical description. But that was not his intention. "Whether or not the heart, besides propelling the blood, giving it motion locally, and distributing it to the body, adds anything else to it—heat, spirit, perfection—must be inquired into by-and-by, and decided on other grounds." He suspected that there was more to the body than physical processes, that the heart was "the sun of the microcosm" and blood a spiritual substance, "the instrument of heaven." But that didn't mean its motions could not be studied systematically.

The words quoted here are from Harvey's masterwork, *On the Motion of the Heart and Blood in Animals.* Though a bit repetitious, the short book, published in 1628 after two decades of research, still makes for a good read. With the tenacity of a prosecutor arguing a case, Harvey marshals his evidence one step at a time. We can imagine him in the courtroom, waving his ceremonial dagger and addressing a jury.

First he asks the audience to consider the arterial system. It was now clear from his experiments that the purpose of the left side of the heart was to pump blood into the arteries, which carried it toward the extremities of the body. It was also clear that unlike the tides this was a one-way flow: there were valves between the left ventricle and the aorta that prevented blood from sloshing back the other way.

Consider next the venous system. It had long been known that veins in the legs and arms contained their own built-in valves. Harvey's teacher in Padua, the great anatomist Fabricius, had discovered these *ostiola,* or "little doors," but thought they served only to slow the blood and prevent gorging. Harvey found the truth by inserting a long probe into a vessel and pushing it in the direction leading away from the heart. The motion was resisted. But it slipped through easily when thrust the other way. The veins were one-way avenues. Arterial blood was pushed out from the heart to the body. Venous blood flowed from the body back to the heart.

Finally consider how the venous blood might get from the right chambers, where it was received, to the left. Harvey had already established that the flow was not through the septum. That left only one possible path—the pulmonary artery connecting the right ventricle to the lungs. It was not air that flowed down the vessel but blood that flowed up the other way. Diffusing somehow through the lungs' spongy tissues, the fluid exited through the pulmonary vein, which led to the left auricle of the heart. The conclusion was inescapable: the right side of the heart pumped blood through the lungs, the left side pumped blood through the body.

Harvey wasn't the first to think of this. In the previous century, a Spanish theologian and physician, Michael Servetus, had speculated about the pulmonary circulation in a religious tract: "Just as by air God makes ruddy the blood, so does Christ cause the Spirit to glow." (His anatomical arguments were part of an attack on the Trinity, and he was ultimately burned—by Protestants—at the stake.) Picking up on the theme, Realdus Columbus, an assistant to Vesalius, noted that the fluid returning from the lungs was bright red, suggesting that vitalization occurred there and not in the heart. It was left for Harvey to ask the crucial question: If the right side of the heart pumps blood through the lungs and into the heart's left side, and if the left side then pumps it out into the arteries . . . then what happens to all the arterial blood when it reaches its destination, and where does the endless supply of venous blood come from?

The Galenists had an answer: both kinds of blood were constantly created from the ingestion of food and depleted in the growth and locomotion of the body. Harvey decided to do the math. From his dissections he had found that the left ventricle is capable of holding two ounces or more of blood, only a portion of which—say, half an ounce—is expelled on each beat. In just one thousand heartbeats (fifteen minutes for an average person) that would come to almost four gallons, far more blood than was present in the entire body. Reckoning by weight instead of volume, the heart would pump well over a ton of blood a day. That would require a lot of eating. And exercise.

So came the radical hypothesis: when blood pumped by the left side of the heart reached the very ends of the arteries, it was picked up by the veins and returned to the right side of the heart. Blood, in other words, moved in a circle. It *circulated.*

He clinched his case with a beautiful experiment.

> If a live snake be laid open, the heart will be seen pulsating quietly, distinctly, for more than an hour, moving like a worm, contracting in its longitudinal dimensions (for it is of an oblong shape), and propelling its contents. It becomes of a paler colour in the systole, of a deeper tint in the diastole.

Using a forceps or thumb and finger, pinch the main vein, the vena cava, just before it enters the heart. The space downstream from the obstruction quickly empties of blood. The heart grows paler and smaller, beating more slowly, "so that it seems at length as if it were about to die." Release the grip and the heart refills with blood and springs back to life.

Next pinch or tie off the main artery just after it leaves the heart. The space upstream from the obstruction is seen to become "inordinately distended, to assume a deep purple or even livid colour, and at length to be so much oppressed with blood that you will believe it about to be choked." Again, when the blockage is removed the heart returns to normal.

Case closed, or so it should have been.

It would be left for others to show with a microscope the tiny capillaries that, in the body's extremities, connected the arteries to the veins, and to explain the osmotic process that carried the blood across the divide. Meanwhile Harvey offered doubters a means of confirming his theory for themselves. Place a tight bandage on your upper arm. Above the bandage, on the side toward the heart, the artery will throb and swell. Below it, toward the hand, there will be no throbbing. At the same time the veins in the lower arm will fill with trapped blood, as the ones above become flaccid. Loosen the bandage slightly, so that it is just tight enough to block off the veins but not the arteries. Then feel the mad rush of blood back to your hand.

Still, hardly anyone believed him. Years later, he was still defending his theory against "detractors, mummers, and writers defiled with abuse." They hounded him like barking dogs, he lamented, "but care can be taken that they do not bite or inoculate their mad humours, or with their dogs' teeth gnaw the bones and foundations of truth."

In 1642, when the English civil war broke out, Harvey, with his royal connections, found himself on the losing side. His home was ransacked and most of his scientific papers

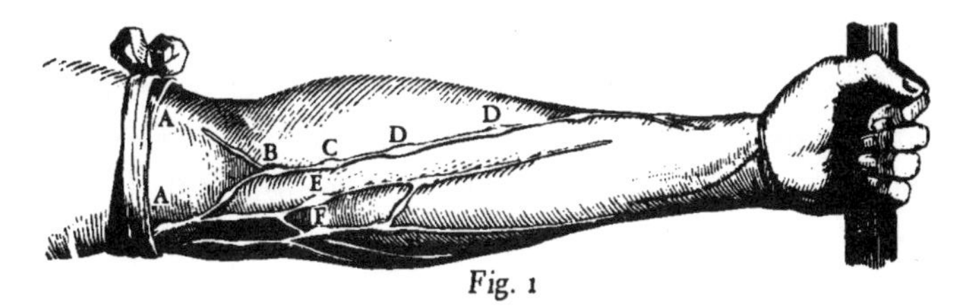

Fig. 1

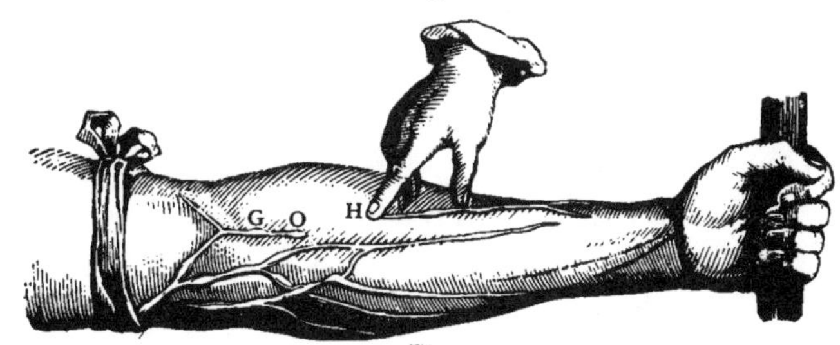

Fig. 2

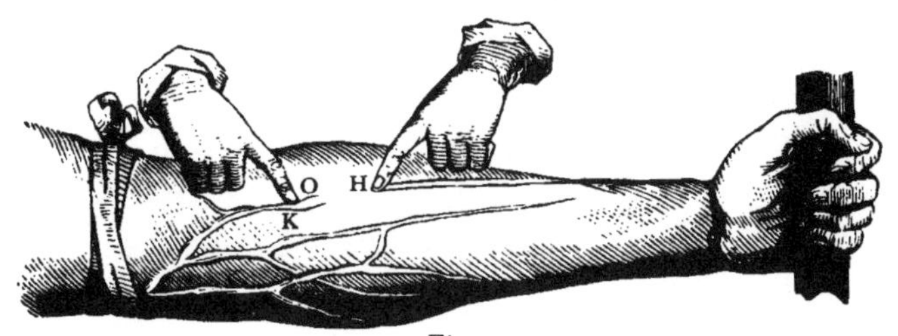

Fig. 3

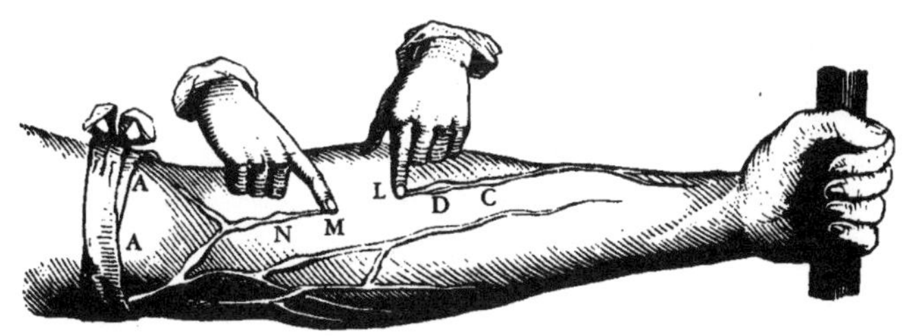

Fig. 4

Blood vessels, from Harvey's *Motion of the Heart*

destroyed. He survived the turmoil, unlike his king, and died fifteen years later, a wealthy man. "But he often sayd, That of all the losses he sustained," his friend Aubrey remembered, "no greife was so crucifying to him as the losse of these papers, which for love or money he could never retrieve or obtaine."

CHAPTER 3

Isaac Newton

What a Color Is

Isaac Newton,
by Sir Godfrey Kneller, 1689

The truth is, the Science of Nature has been already too long made only a work of the *Brain* and the *Fancy:* It is now high time that it should return to the plainness and soundness of *Observations* on *material* and *obvious* things.

—Robert Hooke, *Micrographia*

As you enter the tomb of Isaac Newton, your gaze is swept upward by the vast curved spaces of the vaulted marble ceiling and the massive supporting columns that keep it from succumbing to gravity. Weighing just as heavily is the silence, broken only by the echo of your footsteps ascending the stairs toward the scientist's urn.

It will be then that you notice the light beam. Entering through a tiny hole, perhaps twenty feet above the floor, it shoots down at an angle and ricochets off a mirror mounted on an ornate stand. From there it travels across the room, through a prism, and is transformed into the familiar arpeggio that manifests itself in nature: red, orange, yellow, green, blue, indigo, and violet.

This pantheon exists only in a painting, *An Allegorical Monument to Sir Isaac Newton,* completed by the Venetian artist Giovanni Battista Pittoni in 1729, not long after Newton died. (He is actually buried in Westminster Abbey.) It was something of a departure for Pittoni, who is better known for religious and mythological themes (*The Holy Family, The Sacrifice of Polyxena*). But it was also unusual in another way.

Newton would become known to the ages (along with Leibniz) for his invention of calculus—the "method of fluxions"—which made sense of a concept that had eluded Galileo: how an accelerating object becomes infinitesimally faster during each of an infinity of infinitesimal moments of time. In his later triumph, the *Principia Mathematica,* he described the motions of the heavens and showed that the same gravity that causes an apple to fall holds the planets around the sun. But Pittoni's painting was celebrating some-

An Allegorical Monument to Sir Isaac Newton,
by Giovanni Battista Pittoni

thing different—not Newton the theorist, giver of laws, but Newton the experimenter.

He was barely out of school, having graduated from Trinity College, Cambridge, in 1665, when the Great Plague forced an exodus to the countryside. Trapped at the family farm in Woolsthorpe, he closeted himself in his study, working out some ideas about mathematics and motion and contemplating the peculiarities of color and light.

Plato and some of the Presocratics believed that light beams emanated from the eyes, sweeping the world like searchlights. Aristotle, who rejected that idea, taught that colors are a mixture of light and darkness. Yellow, after all, is nearly white, and blue is almost black. By Newton's time a clearer picture was emerging, and philosophers were developing a precise science of optics.

When light strikes a mirror, they had learned, the angle of incidence equals the angle of reflection. And when it passes through a transparent medium and back into the air, it is bent or refracted—that is why your leg looks broken when you step into a pool of water. The degree of the refraction could be predicted by something that became known as Snell's law. While investigating rainbows, René Descartes, the French philosopher and scientist, had gazed into a giant droplet—a glass sphere filled with water—and studied the colors inside, so much like those that appeared when soap bubbles, flakes of mica, fish scales, and insect wings shimmer in the sunlight. In 1637, in an essay called *Dioptrics,* he tried to account for the origin of color, speculating that it was produced by spinning globules of aether—the faster the rotation, the redder the light.

But no one really knew. Somehow pure white light became stained in its collisions with matter—when it bounced off a colored object or passed through a tinted liquid or piece of glass. A generation after Descartes three of Europe's greatest scientists—Christiaan Huygens, Robert Boyle, and Robert Hooke—were still putting forth theories. None of them had any reason to know about Isaac Newton. Hooke, in particular, would come to wish he had never heard Newton's name.

A stooped troll of a man, Hooke was so well known for his elegant manipulations of nature that he served as the first curator of experiments for the Royal Society of London, which was beginning its emergence as a powerhouse of the scientific revolution. One of the first great microscopists, Hooke produced meticulous drawings—a flea and a louse magnified into monsters, molds as extravagant as flowers in

Viewed under a microscope, "a small white spot of hairy mould." From Robert Hooke, *Micrographia*

a tropical rain forest—that filled the pages of his celebrated book *Micrographia.* Focusing his lenses on a piece of cork, he explored the labyrinth of empty chambers and was the first to call them cells. An ingenious inventor, he designed an air pump and assisted Boyle in discovering the inverse relationship between the volume and pressure of a gas, Boyle's law. There is a Hooke's law as well, precisely describing the nature of elasticity: the amount a solid object can be stretched is proportional to the force that is applied. Or as Hooke himself put it, "ceiiinosssttuv," which unscrambles into *Ut tensio sic vis,* "As the extension, so the force." (To establish priority and avoid intellectual property theft, he first published the law as a Latin anagram.)

Hooke was certain he had also figured out color and light. White was fundamental, and colors were aberrations: "Blue is an impression on the Retina of an oblique and confus'd pulse of light, whose weakest part precedes, and whose strongest follows," he obscurely wrote. Red was the opposite—a misshapen pulse "whose strongest part precedes, and whose weakest follows." Red and blue could be mixed and diluted to form mongrel hues. Huygens and Boyle had their own theories, but they all came down to the same bedrock—color as stained light.

STARTING from scratch, Newton carefully reviewed what others before him had found and added some observations of his own. A piece of gold leaf, thin enough to be almost transparent, reflects yellow light. But hold it "twixt your eye & a candle," he noted, and the light passing through is blue. The opposite effect could be had from a wood called *lignum nephriticum,* sold by druggists as a kidney treatment. When it was sliced into thin pieces and infused in water "the liquor (looked on in a cleare violl) reflects blew rays & transmits yellow ones." The same was true for certain pieces of flat glass: they "appeare of one colour when looked upon & of another colour when looked through." But these were aberrations. "Generally bodys which appeare of any colour to the eye, appeare of the same colour in all positions."

Shut away from the plague, he studied the world with the eyes of a blind man suddenly able to see. Dark or translucent substances ground into a powder or shaved with a knife become lighter in appearance—for the mangling creates a "multitude of reflecting surface" that didn't exist before. By contrast substances soaked in water become darker, "for the water fills up the reflecting pores."

He also played with plates of glass, mounting a flat lens sandwichlike against one with a gentle spherical curve. Shining a light beam at the surface he beheld a mesmerizing pattern of colorful swirls. Newton's rings. "Accordingly as the glasses are pressed more or lesse together the coloured circles doe become greater or less. & as they are pressed more & more together new circles doe arrive in the midst." Taking the apparatus into a dark room he exposed it to a blue ray emitted by a prism. This time he saw a monochromatic target of dark and light circles. Red light produced a similar pattern.

Hooke had already described the phenomenon—interfer-

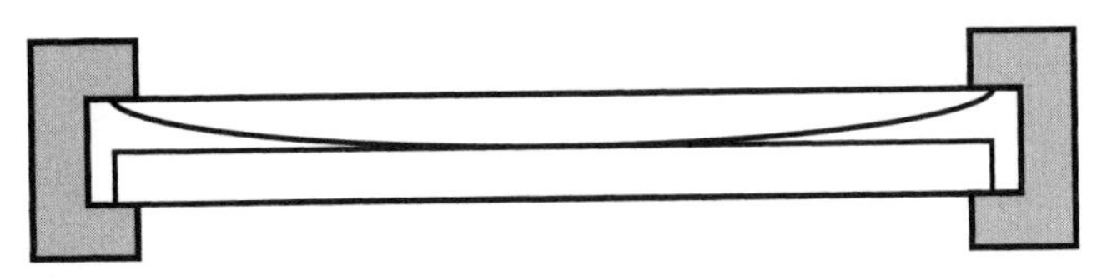

A lens sandwich used to show Newton's rings

ence—in *Micrographia,* but Newton plumbed its depths and made it his own.

As his interests grew into an obsession, he even experimented with his own eyes, taking a thin, blunt probe—a bodkin, he called it—and carefully inserting it "betwixt my eye & the bone as neare to the Backside of my eye as I could." Pressing and rubbing the instrument against his eyeball, he saw "severall white darke & coloured circles." When he repeated the experiment in daylight, with his eyes almost closed, "There appeared a greate broade blewish darke circle" with a smaller, lighter spot inside. If he pressed hard enough, within that spot was another little circle of blue. Performing the experiment in darkness produced a different effect: "the circle apeared of a Reddish light" surrounding an inner circle of "darkish blew."

Sometimes as he poked around in his eye socket he perceived still finer distinctions: a target of colorful rings "from the center greene, blew, purple, darke purple, blew, greene, yellow, red like flame, yellow, greene, blew, broade purple, darke." Staring at the sun or its reflection, he noticed that the afterimage was red, "but if I went into a dark roome the Phantasma was blew."

From physics he occasionally detoured into anatomy. From each eye, he learned in his readings, the visual vibrations traveled through the optic nerves—"a vast multitud of these slender pipes"—and into the brain. Dissecting the tissues around an eye—an animal's, thank God, not his own—he tried to determine the nature of the substance that carried the imagery. "Water is too grosse for such subtile impressions," he concluded. A better possibility seemed to be the "animal spirits" said by the Galenists to blow through the nervous system. Newton ruled that out with an experiment:

paper, wood, marble, ye Oculus Mundi Stone, &c) become
more darke & transparent by being soaked in water
[for ye water fills up ye reflecting pores]
58 ~~If with a bodkin gh~~
58 I tooke a bodkine gh
& put it betwixt my
eye & ye bone as
neare to ye ~~end of~~
backside of my eye
as I could: & pressing
my eye wth ye end of
it (soe as to make ye
curvature a, bcdef in my
eye) there appeared severall
white darke & coloured circles
r, s, t, &c. Which circles were
plainest when I continued to rub my eye wth ye
point of ye bodkine, but if I held my eye & ye
bodkin still, though I continued to presse my eye
wth it yet ye ~~colours~~ circles would grow faint
& often disappeare untill I renewed ym by moving
my eye or ye bodkin.
59 If ye experiment were done in a light roome so
yt though my eyes were shut some light would
get through their lidds There appeared a ~~[illegible]~~
~~or redish spot in ye midst at s~~ greate broade
blewish darke circle outmost (as ts), & wthin that
another light spot srs whose colour was much
like yt in ye rest of ye eye as at k. Within
wch spot appeared still another blew spot r

Newton's experiment with his own eye: a page from his notebooks

"though I tyed a peice of the optick nerve at one end & warmed it in the middle to see if any aery substance by that meanes would disclose it selfe in bubbles at the other end, I could not spy the least bubble; a little moisture only & the marrow it selfe squeezed out."

If that is where it all had ended—waiting for the spirits of vision to come bubbling from the optic tubules—Newton might have remained just another seventeenth-century genius confused and tantalized by light. But sometime in the midst of his investigations he became captivated by a curiosity involving prisms. Draw a line, half blue and half a "good deepe red," on a black piece of paper and the prism will make it appear skewed: "broken in two twixt the colours." The same thing happened with blue and red threads. One was offset from the other. But why were the colors treated differently by the glass?

One day, his curiosity aroused, he cut a small circular hole a quarter-inch across in his window shutter. Holding a prism in the narrow path of the sunbeam, he cast a spectrum on the far wall of the darkened room.

"It was at first a very pleasing divertisement to view the vivid & intense colours," he reported: blues fading into greens then yellows into oranges and reds. But far more significant than the familiar appearance of a spectrum was its

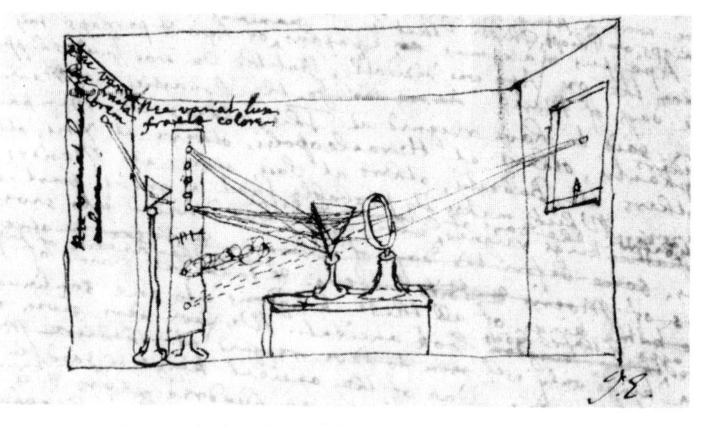

Newton's drawing of his *Experimentum Crucis*

shape. It was not circular like the hole in the shutter or the image of the sun, but oblong: thirteen and one-fourth inches long, two and five-eighths inches wide. It was "a disproportion soe extravagant that it excited me to a more then ordinary curiosity of examining from whence it might proceed."

Something was causing the colors to fan out this way. Newton doubted that the effect could be an artifact, some obscure confluence of accidental effects. But the possibility had to be ruled out. He tried holding the prism in different positions so that the light traveled "through parts of the glasse of divers thicknesses." He cut holes in the shade of "divers bignesses." He tried putting the prism outside the window, so the sunlight hit it before passing through the hole. None of that mattered. "The fashion of the colours was in all these cases the same."

Having refracted sunlight with one prism, he found that he could pass the colors through a second prism and they would recombine. The second prism undid what the first had done, leaving a colorless circle of light on the wall. The colors were not added by the prism. They had been in the light beam all along.

It was a multitude of such experiments that led him to his surprising conclusion. By the time he was ready for what he would call his *Experimentum Crucis* (borrowing the term from Hooke), he probably knew what he would find. But that barely detracts from the drama. As before, the light beam from the window passed through a prism and crossed the room, but this time it cast its spectrum on a wooden board. Through one end of the board Newton had drilled a hole, and by holding his prism just so, he could make the colors pass through the opening one by one. From there they entered a second prism before leaving an image on the wall.

What he saw that day changed forever how we think about light. Starting at the red end of the spectrum and progressing toward the blue, each color was bent a little more—an elaboration of the effect hinted at by the colored threads: "blew rays suffer a greater refraction than red ones." That was the reason for the oblong. If all colors were bent equally the spectrum would be a roundish blob. But light, as Newton put it, "consists of rayes differently refrangible."

Refrangible means refractable—both words come from the same Latin root—and Newton had discovered nothing less than what a color is: a ray of light preternaturally disposed to bend a certain way. "To the same degree of refrangibility ever belongs the same colour, & to the same colour ever belongs the same degree of refrangibility," he wrote. Color *is* refrangibility.

And there was more. Once a color was separated from the rest, it could not be further altered, no matter how hard he tried. "I have refracted it with Prismes, & reflected it with bodies which in day light were of other colours; I have intercepted it with the coloured film of air interceding two compressed plates of Glasse, transmitted it through coloured mediums & through mediums irradiated with other sort of rayes, & diversly terminated it, & yet could not produce any new colour out of it. It would by contracting or dilating become more brisk or faint, & by the losse of many rayes in some cases very obscure & dark, but I could never see it changed in Specie."

If a ray was composed of more than one color—orangish yellow, yellowish green—it could be split once again by a prism, but at some point you would reach the bottom, the fundamental components of light. "Colours are not qualifications of light derived from refractions or reflections of naturall bodies as 'tis generally beleived, but originall & connate properties."

It was white light that was the mongrel, not just another color but a combination of them all, a "heterogeneous mixture of differently refrangible rayes." As the sun shines on the world, it is not bringing out the red in an apple, the green in a leaf. The apple and the leaf are bringing the colors out of the sunlight.

Descartes had also believed that colors were not inherent in objects, but rather manifestations of how they affected light. Now Newton knew why. The world is colorful because it consists of bodies "variously qualified to reflect one sort of light in greater plenty than another."

IN EARLY September 1666, the Great Fire destroyed much of London, killing the rats and hastening the end of the plague. Setting aside optics and other scientific pursuits, Robert Hooke worked with Christopher Wren to rebuild the city. Newton moved back to Cambridge, where he rose to the position of Lucasian professor of mathematics and lectured on color and light. A reflecting telescope he invented, six inches long and more powerful than a conventional telescope ten times its size, impressed the members of the Royal Society, and in 1672, six years after his experiments, they published his paper "New Theory About Light and Colors" in the society's *Philosophical Transactions.*

Burning with jealousy, Hooke tried to discredit the upstart, setting off a feud that would last as long as both men were alive. Hooke declared that he had already performed all these experiments himself, and that the results could be explained just as well by his own theory. (Later he would claim that Newton's *Principia* was plagiarized from him.)

Other scientists, like Huygens, also raised objections in dispatches to the journal, and Newton countered his naysayers with a mixture of disbelief and scorn. The merciless dissection of new ideas would become a normal part of science. But Newton, an intensely private man, felt violated. He became especially agitated by a group of English Jesuits who insisted that they could not replicate his *Experimentum Crucis* and that the spreading out of the spectrum was an artifact caused by a "bright cloud." The carping continued until 1678, when in exasperation he retreated into seclusion. He was thirty-five. There was so much still to be done.

CHAPTER 4

Antoine-Laurent Lavoisier

The Farmer's Daughter

Antoine-Laurent Lavoisier

> Imagine what it means to understand what gives a leaf its color! What makes a flame burn.
>
> —Marie Anne Lavoisier in the play *Oxygen*, by Carl Djerassi and Roald Hoffmann

OUTSIDE the Louvre in the Jardin de l'Infante on an autumn day in 1772, Parisians strolling along the Seine might have noticed a strange contraption: a wooden platform on six wheels, like a flatbed wagon, on top of which was mounted an assembly of enormous pieces of glass. The two largest lenses—eight feet in radius—had been sandwiched into a single powerful magnifier that captured the solar rays, beaming them through a second, smaller lens and onto a table. Standing on deck, scientists in wigs and dark glasses were performing an experiment, while assistants, like midshipmen, cranked gears and adjusted the rigging, following the sun across the sky.

One of the men who had booked time on the machine—the particle accelerator of its day—was Antoine-Laurent Lavoisier. He was trying to find out what happens when you incinerate diamonds.

It had long been known that diamonds burn (we now know that they are made of carbon), and local jewelers had asked the French Academy of Science to investigate whether this posed a risk. Lavoisier himself was more interested in another question: the chemical nature of combustion. The

Incinerating diamonds

beauty of the "burning glass" was that it could focus sunlight onto a spot inside a closed container, heating whatever had been placed inside. The fumes from the jar could then be channeled through a tube into a flask of water, gurgling up to form a bubble to be drawn off and analyzed.

The experiment was a failure: the intense heat kept cracking the glass. But there were other items on Lavoisier's agenda. What he had proposed to the Academy of Science was a program to study "the air contained in matter" and how it might be related to the true nature of fire.

ALTHOUGH Newton had put physics on a straighter path, he hadn't been much help with chemistry, which was still in the thrall of alchemy. "Camphire dissolved in well deflegmed spirit of niter will make a colourlesse solution," he had written. "But if it bee cast into good Oyle of Vitriol & shaken into it as it dissolves, the liquor will bee first yellow & then of a deepe reddish colour." In page after page of this cookbook chemistry, there was little talk of measurement or quantification: "Putting spirit of salt to fresh urin the two liquors readily & quietly mix," he noted, while "if the same spirit be dropped upon digested urin there will presently ensue a hissing & ebullition, & the volatile & acid salts will after a while coagulate into a third substance, somewhat of the nature of Salarmoniac. And whereas the syrup of Violets is but diluted by being dissolved in a little fresh urin, a few drops of fermented urin presently turns it into a deep green."

That, at least, was protochemistry. Much of alchemy, including Newton's own, sounds to modern ears like magic. In one of his notebooks, he had dutifully copied passages from an alchemist named George Starkey, who called himself Philalethes.

"In [Saturn] is hid an immortal soul," the passage began. Saturn usually meant lead—each element resonated with a planet—but here it apparently refers to a silvery metal called antimony. The "immortal soul" is the gas emitted when the ore is exposed to an intense flame. "To Saturn Mars with bonds of love is tied"—iron is added to antimony—"who is by him devourd of mighty force whose spirit divides saturns body & from both combined flow a wondrous bright water in which ye Sun doth set & loos its light." The sun is metallic gold, and here it is immersed in mercury, commonly called quicksilver. "Venus a most shining star is embract'd by [Mars]." Venus was copper. Now it too was added to the mix. The metallurgical recipe is apparently a description of the early stages of producing the long-sought "philosopher's stone," capable of transmuting baser elements into gold.

Lavoisier and his peers had moved beyond such mystical incantations, but chemists still commonly accepted the alchemical notion that matter was governed by three principles: mercury (which made things liquid), salt (which made them solid), and sulfur (which made them burn). The sulfurous spirit, also called *terra pingua* ("fat" or "oily" earth) was a special source of fascination. Early in the 1700s, a German chemist, Georg Ernst Stahl, renamed it phlogiston from the Greek root, *phlog,* referring to fire.

The reason things burned was that they were rich in phlogiston, and as they were consumed they released this fire stuff into the air. Set a piece of wood aflame and it would stop burning only when its phlogiston was spent, leaving behind a pile of ash. Wood, it logically followed, was made of phlogiston and ash. Likewise, heating a metal under an intense flame, a process called calcination, left a whitish brittle substance, or calx. Metal was thus composed of phlogiston and calx. Rusting was another form of this slow combustion, as was respiration—reactions caused when phlogiston is given up to the air.

The process also worked the other way around. Calx, it was recognized, resembled the crude ores mined from the ground, which were refined or *reduced*—"revivified"—by heating them next to a piece of charcoal. The charcoal emitted phlogiston, which combined with the calx to recover the lustrous metal.

There was nothing necessarily wrong with invoking a hypothetical entity that could not be measured but only inferred. In our own time, cosmologists propose that an intangible "dark matter" must exist to keep the galaxies from spinning apart from their own centrifugal forces, and that an antigravitational "dark energy" propels the cosmological expansion.

With phlogiston, scientists had a consistent explanation for combustion, calcination, reduction, and even respiration. Chemistry suddenly made sense.

There was however a problem: the calx left behind after calcination weighed more than the original metal. How could removing phlogiston leave something heavier? Like dark energy a quarter of a millennium later, phlogiston was, in the words of the French philosopher Condorcet, "impelled by forces that give it a direction contrary to that of gravity." Putting it more poetically, one chemist declared that phlogiston "gave wings to earthly molecules."

Lavoisier too had learned to think of phlogiston as one of the principal ingredients of matter. But around the time of his experiments with diamonds he was beginning to wonder: How could something weigh less than zero?

HIS MOTHER had died when he was a boy, leaving an inheritance large enough for him to buy into a profitable enterprise called the Ferme Générale, or General Farm. The French government contracted with this private consortium of businessmen to collect certain taxes, giving the "farmers," like Lavoisier, a cut. Though his duties took time away from research, he made enough money to equip himself with one of the best laboratories in Europe. One of his early experiments, in 1769, investigated the commonly held belief that water could turn into earth.

The evidence seemed persuasive: water evaporating in a pan leaves behind a solid residue. Lavoisier cut to the heart of the matter with a glass distilling flask called a pelican. Round and fat at the base with a small upper chamber, the vessel was outfitted with two curving tubes (shaped a bit like pelican

A pelican flask. John French, *The Art of Distillation* (London, 1651)

beaks) that returned condensed vapors to the bottom. To the alchemists, the pelican symbolized the sacrificial blood of Christ, and a pelican flask was said to have transformative powers. More to the point, water boiled in a pelican would continually evaporate and recondense without anything—solid, liquid, or gas—leaving the system.

After distilling pure water for a hundred days, Lavoisier found that a residue had indeed accumulated. But he suspected where it had come from. Weighing the empty pelican, he confirmed that it was lighter than before. When he dried and weighed the leftover debris, it matched closely enough to convince him that it had come from the glass.

Two years later, in 1771, Lavoisier, who was then twenty-eight, married Marie Anne Pierrette Paulze, the thirteen-

Marie Anne Pierrette Paulze

year-old daughter of another tax farmer. (She was pleased enough with the arrangement: her other suitor was fifty years old.) Fascinated by her husband's research, Marie Anne learned chemistry at his side, recording notes, translating English scientific literature into French, and producing the meticulous drawings for a series of experiments crowned by one so beautiful that—like a philosopher's stone—it transformed alchemy into chemistry.

THE CHEMISTS of Lavoisier's generation had already discovered that there were, as the Englishman Joseph Priestley put it, "different kinds of air." Mephitic (meaning noxious), or "fixed air," would extinguish a flame and suffocate a mouse. It also turned limewater (calcium hydroxide in modern terms) cloudy, forming a white precipitate (calcium carbonate). But plants thrived in the gas and slowly made it breathable again.

There was another suffocating gas left behind when a candle was burned in a covered jar. It did not precipitate limewater, and since it was evidently related to combustion was called "phlogisticated" air—or *azote,* from the Greek word for "lifeless." Most mysterious of all was a volatile gas emitted when iron filings were dissolved in dilute sulfuric acid. It was so combustible that it was named "inflammable air." A balloon filled with it would float high above the ground.

The question was whether these new airs were elements or, as Priestley believed, modifications of "normal" air, produced by adding or subtracting phlogiston.

Carefully keeping his skepticism in check, Lavoisier repeated some of his colleagues' work. He confirmed that burning phosphorus to make phosphoric acid or sulfur to make sulfuric acid indeed left the substances heavier—the same thing that happened when you calcined metals. But what was causing the change? He thought he knew the answer. Using a burning glass to heat tin that had been sealed inside a flask, he found that the entire apparatus weighed the same before and after. Slowly opening the vessel he heard air whistle in, and only then was there a gain in weight. Maybe things burned not because they emitted phlogiston but because they absorbed some kind of air.

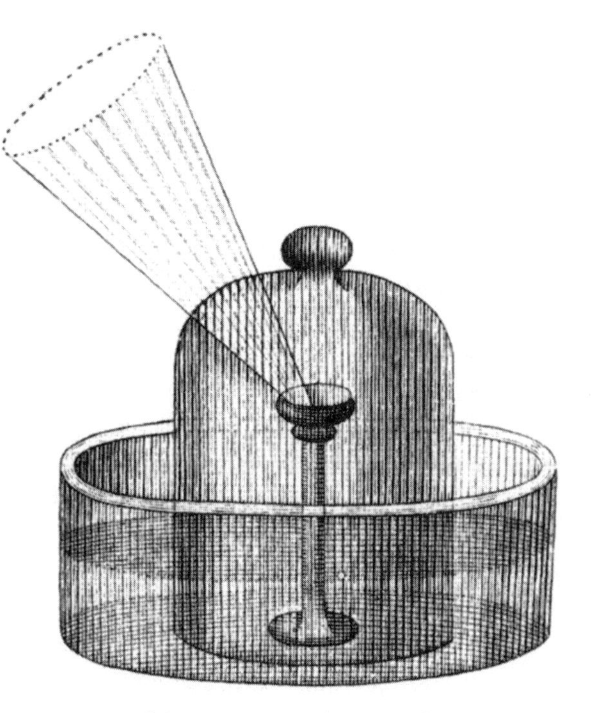

Burning litharge in a jar with a magnifying glass.
Drawing by Marie Anne Lavoisier

If so, then reducing a substance—smelting an ore back into a pure metal—should force the air back out again. He measured out a calx of lead called litharge and placed it, along with a piece of charcoal, on an island pedestal in a basin of water. Then he inverted a bell jar on top. Heating the calx with a magnifying glass, he could see from the displacement of the water that a gas was coming out. Carefully withdrawing it from the jar, he found that it extinguished flames and precipitated limewater. Fixed air appeared to be a product of reduction, but was there more to the story than that?

The answer turned out to lie in a reddish substance called *mercurius calcinatus,* or calx of mercury, sold by Parisian apothecaries as a treatment for syphilis. With a price of eighteen *livres* and up per ounce—about $1,000 in today's currency—experimenting with *mercurius calcinatus* was almost as extravagant as burning diamonds. Like all calxes, it could be produced by heating the pure metal over an intense flame. But when heated further it changed, against all expectations, back into quicksilver. In other words, *mercurius calcinatus* could be reduced without the presence of charcoal. But what then was supplying the phlogiston? In 1774, Lavoisier and some colleagues from the French Academy confirmed that calx of mercury could indeed be reduced "without addition," losing about one-twelfth of its weight.

Priestley was also experimenting with the stuff, heating it with a magnifying glass and collecting the fumes. "What surprised me more than I can well express," he would later report, "was that a candle burned in this air with a remarkably vigorous flame. . . . I was utterly at a loss how to account for it." After finding that a laboratory mouse thrived on the gas, he tried breathing it himself. "I fancied that my breast felt peculiarly light and easy for some time afterwards. Who can tell but that, in time, this pure air may become a fashionable article in luxury. Hitherto, only two mice and myself have had the privilege of breathing it."

A gas in which the fires of combustion and respiration flourished must be a particularly good absorber of phlogiston, so Priestley named it "dephlogisticated air"—air in its very purest form. He was not the only one thinking along this line. In Sweden, an apothecary named Carl Wilhelm Scheele was studying the properties of what he referred to as "fire air."

By now Lavoisier was calling the gas expelled by reducing *mercurius calcinatus* "eminently breathable" or "vital" air, and like Priestley he thought it was ordinary air in its pristine form. But he had run across a complication. When he tried reducing mercury calx with charcoal—the old-fashioned way—it released the same gas he had obtained from litharge: one that extinguished candles and precipitated limewater. Why would reducing mercury calx without charcoal produce vital air while reducing it with charcoal produced suffocating fixed air?

There was one way to find out. He took from his shelves a flask called a matrass, round on the bottom with a long skinny neck, which he heated and bent so that it curved down and then up again.

If the flask in his experiment of 1769 resembled a pelican, this one looked more like a flamingo. He poured four ounces of pure mercury into the round bottom chamber (A on the diagram) and set it on a furnace with the neck dipping down into an open trough, also filled with mercury, and then up into a bell jar. This would serve as a gauge to measure how

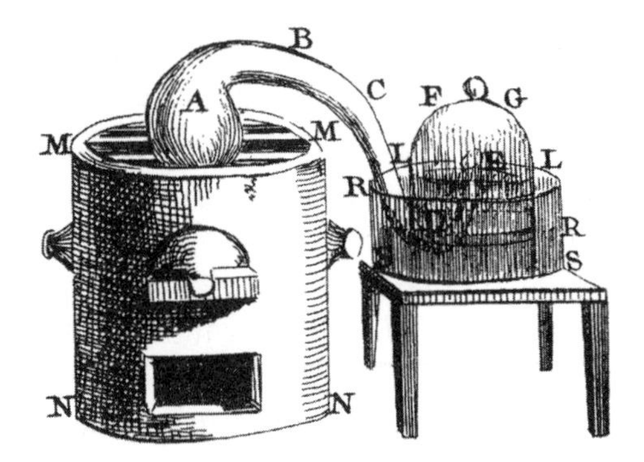

Heating mercury in a "flamingo flask."
Drawing by Marie Anne Lavoisier

much air was consumed during the experiment. After marking the level (L L) with a paper label, he lit the furnace and brought the liquid metal in chamber A almost to a boil.

On the first day nothing much happened. Small amounts of quicksilver evaporated and condensed along the wall of the matrass, combining into blobs heavy enough to slide back down to the bottom. But on the second day tiny red spots started to appear on the surface of the mercury—the calx. For the next few days the reddish crust increased in size until it could grow no larger. On the twelfth day Lavoisier stopped the experiment and took some measurements.

By now the mercury in the bell jar had risen above the mark, displacing some of the air that had been absorbed by the calx. Adjusting for temperature and pressure changes in the lab, Lavoisier calculated that the air had been depleted by about one-sixth of its volume, from fifty cubic inches to between forty-two and forty-three. It had also changed in nature. When a mouse was put inside a container of this leftover air, it struggled for breath, and "when a taper was plunged into it, it was extinguished as if it had been immersed into water." But since the gas did not precipitate limewater it had to be *azote* rather than fixed air.

But what had the burning mercury taken from the air? Skimming off the red crust that had formed on the metal, Lavoisier heated it in a retort until it turned back into pure quicksilver, emitting seven or eight cubic inches of gas—approximately the same amount that had been absorbed during calcination. Exposed to this gas a taper burned "with a dazzling splendor" and charcoal, rather than smoldering, "threw out such a brilliant light that the eyes could hardly endure it."

It was a pivotal moment. Burning mercury absorbed vital air from the atmosphere, leaving behind *azote.* Reducing mercury released the vital air again. He had separated the two main components of the atmosphere.

In a denouement he recombined eight parts of his vital air with forty-two parts of the *azote* and showed that it had the characteristics of ordinary air. Analysis and synthesis: "Here is the most complete kind of proof that can be attained in chemistry, the decomposition of air followed by its recomposition."

Lavoisier read the results to the Academy of Science in 1777. There was no phlogiston. Burning and calcination were caused when a substance took in vital air—oxygen he would call it because of its role in the formation of acids. (*Oxy* in Greek means sharp.) When the oxygen is depleted from the air by burning, the unbreathable *azote* left behind is nitrogen.

As for the gas that people had been calling fixed air, it was produced when the oxygen emitted during reduction combined with something in the charcoal, producing what we now call carbon dioxide.

FOR YEARS Lavoisier's colleagues, particularly Priestley, grumbled that he had grabbed credit for work they had also done. Priestley had dined with the Lavoisiers, telling them about his dephlogisticated air, and Scheele, the Swedish apothecary, had sent Lavoisier a letter describing his work. But all the while they continued to think of oxygen as air devoid of phlogiston.

In their play *Oxygen,* which premiered in 2001, two chemists, Carl Djerassi and Roald Hoffmann, imagine the three scientists summoned to Stockholm by the king of Sweden to decide who should be revered as the true discoverer. Scheele was the first to extract the gas and Priestley the first to publish word of its existence, but only Lavoisier understood what he had found.

He had also seen through to something deeper: the law of conservation of mass. In a chemical reaction, matter—the burning mercury, the altered air—changes form. But mass is neither created nor destroyed. The same amount going into the transaction must come out the other end. The ledgers must balance, a tax collector would say.

In 1794, during the Reign of Terror, Lavoisier and Marie Anne's father were convicted along with other tax farmers as enemies of the state and brought by wagon to the Place de la Révolution, where a wooden platform had been erected, every bit as imposing as the one on which Lavoisier had once burned diamonds. In place of the giant lenses was another example of French technology, the guillotine.

A story ricocheting through the Internet a while ago insisted that before his execution Lavoisier arranged to perform one final experiment. The guillotine had been promoted in France as a particularly humane form of execution, manufacturing instantaneous and painless death. Here was a chance to find out. The moment he felt the blade touch his neck Lavoisier would begin blinking his eyes as many times as he could. An assistant standing in the crowd would count the blinks. The story is probably not true. But it sounds like just the kind of thing Lavoisier might have done.

CHAPTER 5

Luigi Galvani

Animal Electricity

Luigi Galvani

> For it is easy in experimentation to be deceived, and to think one has seen and discovered what we desire to see and discover.
>
> —Luigi Galvani

MIDWAY through the eighteenth century, when electricity was all the rage, an amateur scientist stood before the Royal Society in London and described what might be called Symmer's law: opposite-colored socks attract while like-colored socks repel. To keep his feet comfortable in winter, the speaker, a government clerk named Robert

Symmer, was accustomed to wearing two layers of stockings. In the morning he would pull white silk socks over a black woolen pair. In the afternoon he would reverse them. During the transition, the two different materials crackled and bristled with opposite charges, and Symmer, who became known as the barefoot philosopher, would sit back in his chair marveling at the results.

"When this experiment is performed with two black stockings in one hand, and two white in the other," he reported, "it exhibits a very curious spectacle: The repulsion of those of the same colour, and the attraction of those of different colours, throws them into an agitation that is not unentertaining."

This was the height of the romantic era in electrical research, with scientists debating whether electricity was a vapor, a fluid, or, as Benjamin Franklin speculated, "subtle particles." Cranking the wheels of their static-electricity generators—great spinning disks and globes that were rubbed to produce a charge—scientist-entertainers (they were called "electricians") sent shock waves traveling hand by hand through human chains. Suspend a man in a chair with silk ropes (to keep him from being grounded) and his head could be made to glow like the gold leaf aura around the image of a saint. A young woman, picked from the audience and given a charge, would electrify her suitor with an unforgettable kiss. Positive, meet negative.

Ghostly as it seemed, electricity was tangible enough to store in a jar. Wrapped inside and out with two pieces of foil connected to opposite poles of a friction generator, the vessel took on a charge—negative on one side of the glass and positive on the other—that lingered long after the wires were removed. Touching both sides of this primitive capacitor, called a Leyden jar, was like being stung by an eel.

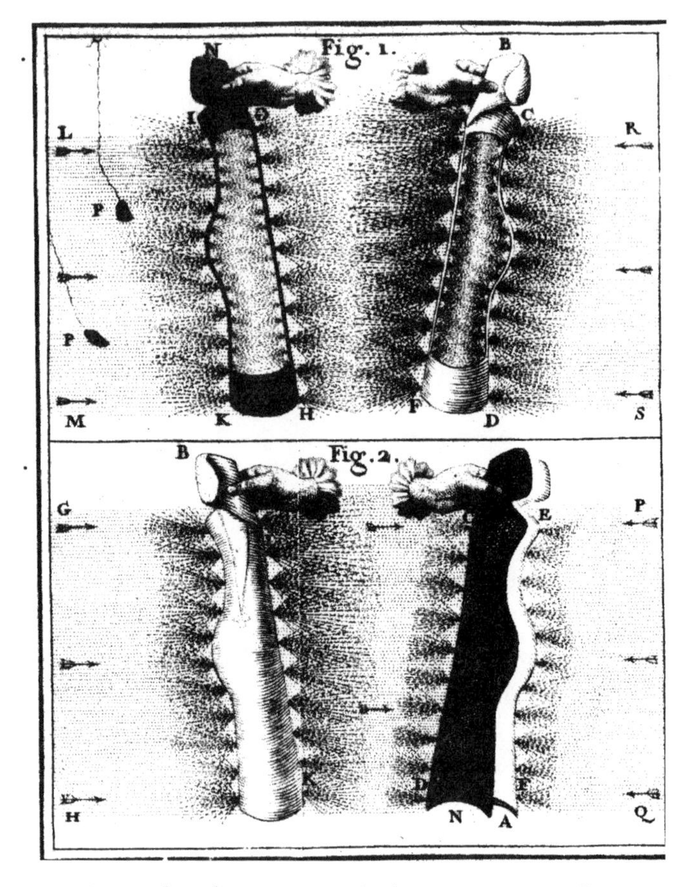

Symmer's socks. From a treatise by Jean-Antoine Nollet, a French abbé and physicist

Empirical fact tangled with fantasy as scientists deliberated over reports of lightning spontaneously causing cripples to walk or plants to grow faster. Speculating that electricity was produced in the brain—from the conversion of phlogiston—Joseph Priestley went on to propose that it was respon-

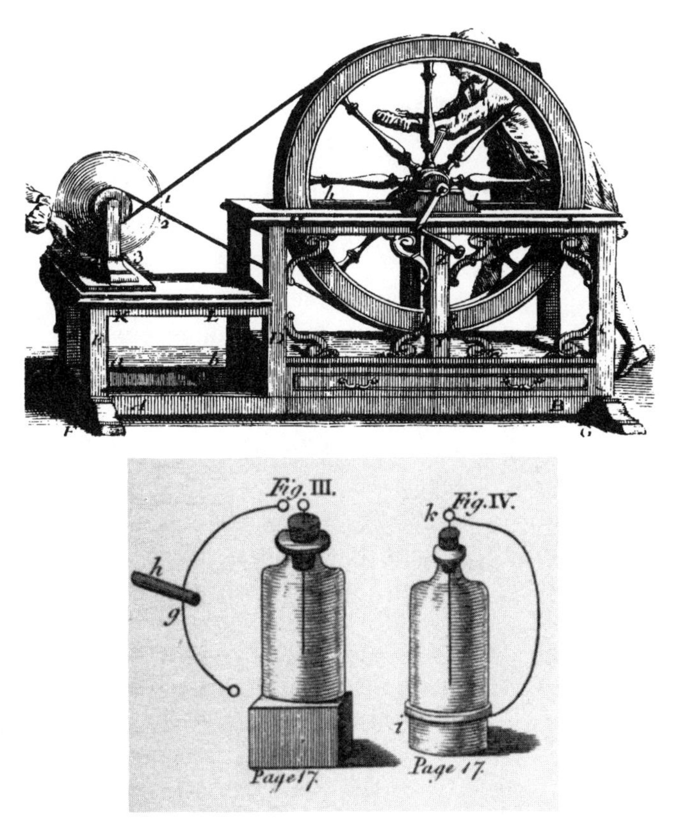

An eighteenth-century static electricity machine and
Benjamin Franklin's drawing of two Leyden jars

sible for muscular motion . . . as well as for the iridescent sheen of parakeet feathers and the light "said to proceed from some animals" when they stalked their prey at night, and even from people "of a particular temperament, and especially on some extraordinary occasions."

Others thought some kind of "nerveo-electrical" fluid was produced in the body by friction. It was a startling idea. Like

Symmer's socks, nerves and bones would rub against muscles, generating the life force, electricity.

ON AN April evening in 1786, more than a quarter century after Symmer's discovery, Luigi Galvani, a middle-aged professor of anatomy, walked to a terrace at the Palazzo Zamboni near his home in Bologna, carrying a roll of metal wire and the legs of a frog prepared, as he often put it, "in the usual manner": severed at the spinal cord with the sciatic (or crural) nerves dangling out.

As clouds gathered to the south, he positioned the headless specimen on a table and connected it to a clothesline of wire, which he had strung overhead. Then he waited for an electrical storm, observing that the legs twitched in response to lightning as though warning of the coming thunder.

TABLE II

Muscular contractions caused by lightning. From Galvani's *De Viribus Electricitatis in Motu Musculari Commentarius*

Over the years Galvani had produced similar effects in his laboratory, stimulating frog nerves with electricity cranked from a generator or discharged from a Leyden jar. The demonstration above the Palazzo Zamboni confirmed for him that "natural" electricity produced the same physiological reaction as "artificial" electricity. One way or another, it made muscles move.

There was one experiment, however, that he was finding harder to interpret. Several years earlier one of his assistants had happened to touch a scalpel to a frog's exposed nerve just as a second assistant, working nearby with a generator, created a small spark. No wires ran from the machine to the dissected animal, but its legs contracted violently, as if in a seizure. Galvani had been investigating the phenomenon ever since.

Early on he established that the response wasn't caused simply by irritation from the scalpel. Making sure the gener-

TABLE I

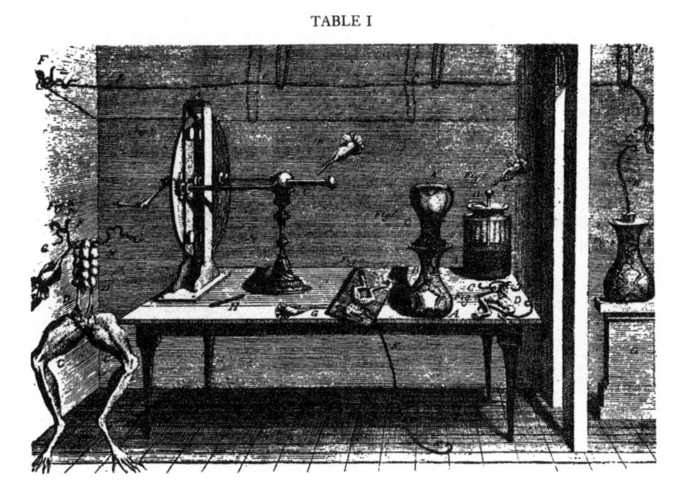

Static electricity and frogs' legs

ator was idle, he pressed against the nerve with a metal blade. No matter how insistently he probed, the muscles lay motionless. The effect clearly appeared to be electrical.

Other experiments showed that an iron cylinder would pick up the spark and make the legs twitch, but not a glass rod. Sometimes, however, even a metal scalpel failed to provoke a response. Galvani quickly realized that these failures occurred when he held the instrument by its bone handle without touching the rivets or the blade. Somehow the experimenter himself seemed to be part of the reaction. To test this hypothesis, he placed the metal cylinder by itself on the table so that it touched the nerve, and then cranked the generator. The leg lay still.

Step by step, he eliminated the variables. If he connected the nerve to a long metal wire instead of a short cylinder, a distant spark did make the legs jump. The situation was becoming a bit clearer. Scientists already knew that electricity could exert its influence across a distance: the hairs on a human neck bristled when a lightning bolt struck nearby. Cranking the generator caused a tension to build in the air—an "electrical atmosphere." The holder of the scalpel and the scalpel itself served as a kind of antenna—a lightning rod—discharging itself through the frog.

But maybe, Galvani suspected, something even stranger was happening. If the frog was merely reacting to artificial electricity transmitted through the air, the intensity of the twitching should depend on the proximity of the spark. Attaching a metal hook to a frog's spinal cord and the hook to a length of wire, he repeated the experiment at various distances, placing the frog as far as 150 feet away from the generator. The reaction was as vigorous as ever—even when the legs were shielded inside a tin cylinder or isolated in a vacuum chamber. One variation after another seemed to point to what Galvani had instinctively come to believe: that the electricity produced by the machine was not the primary cause of the jumping. It was no more than a trigger, exciting a naturally occurring "animal electricity" that flowed through the nerves.

Galvani knew how easy it was for an experimenter to fool himself, to see what he wanted to see. Warily he circled his prey. Early in September, several months after the experiment at the Palazzo Zamboni, he took several of his truncated frogs and hung them by metal hooks from an iron railing on his balcony. This time there was no lightning, no generators sparking, and yet the legs twitched anyway.

The electricity could not be originating inside the metal, he reasoned. A single conductor—the hook and rail—cannot hold a charge. To create a potential, negative and positive must be kept carefully apart, as in a Leyden jar. Harder to discount was the possibility that atmospheric electricity had somehow "crept into the animal and accumulated," rushing out when the hook made contact with the rail. The sky that day was clear, but Galvani wanted to rule out the possibility.

With one hand he picked up a frog, dangling it by its implanted hook so that the feet touched the top of a silver box. Holding a piece of metal in his other hand, he touched it to the same shiny surface, completing a circuit and causing the frog to jump. The same thing happened when he held the frog by the torso so that both the hook and one of the feet brushed against the flat conductor: "At the very moment the foot touched the surface, all the leg muscles contracted, lifting the leg." When the foot fell back to the surface, it contracted again . . . and again, the frog hopping and hopping until its energy was spent. What could this be but animal electricity?

In 1791, Galvani published his findings as *De Viribus Electricitatis in Motu Musculari Commentarius* (*Commentary on the Effect of Electricity on Muscular Motion*), proposing that the frog's muscle was like a Leyden jar, storing and discharging some kind of organic electricity. After carefully describing his experiments and analyzing the results, he allowed himself to speculate. In people, he proposed, an excess of electricity might cause fidgeting, flushing, or in extreme situations epileptic seizures. Venturing briefly outside his own area of expertise, he suggested that lightning and earthquakes might somehow be related: "But let there be a limit to conjectures!" In time he hoped to investigate whether electricity was involved in all manner of bodily functions: "on circulation of the blood and secretion of the humors, these things we will publish as soon as possible in another commentary, when we have found a little more leisure."

AT FIRST Alessandro Volta, one of Europe's greatest electricians, was impressed by Galvani's discovery, declaring that the experiments had placed animal electricity "among the demonstrated truths." Then he politely proceeded to dismantle the theory piece by piece.

Taking as his subject an entire frog, he tried touching its back with a strip of metal and its leg with a coin or a key. Then he closed the arc by bringing the tops of the two probes together. The result was "the same convulsions, spasms and jerks" that Galvani had reported—but only if he used two different kinds of metals.

Galvani had reported in his own experiments that a "bimetallic arc" seemed to amplify the contractions, but he considered this no more than a diverting detail. At first Volta was similarly inclined, proposing that the combination of metals somehow encouraged the flow of the frog's own electricity as it rushed through the completed circuit. But then he took a closer look.

After exposing a sciatic nerve, he attached two tiny metal clips, like collars, leaving a slight gap in between. One clip was tin and the other was silver. The moment he closed the circuit—touching the clips together or bridging them with a wire—the limb convulsed. He produced a similar effect with tin and brass. The conducting arc, Volta was coming to believe, was not just a quiescent connection discharging or even accelerating animal electricity. It was the actual source of the energy. When the frog's leg twitched, it was acting like the needle of a very sensitive meter. What it was indicating was the presence of a newly discovered phenomenon: bimetallic electricity. "Galvani's theory and explanations . . . are largely disqualified," Volta wrote to a colleague, "and the entire edifice is in danger of collapsing."

When Galvani's frog danced on the lid of a silver box, it was merely reacting to electrical shocks. Volta's conclusion was as gentlemanly as it was cruel: "If that is how things are, what is left of the animal Electricity claimed by Galvani, and seemingly demonstrated by his very fine experiments?"

GALVANI was quick to rise to the challenge. It was true that brass hooks had been used to hang frog legs from an iron rail. But the arc did not have to be bimetallic: he reported similar results with iron hooks. Returning to the laboratory, he and his supporters showed that they could elicit convulsions by simultaneously touching muscle and nerve with two pieces of metal that were obviously identical.

Volta was ready with an answer. A piece of metal may seem to be homogeneous, but inevitably there would be impurities—imperceptible differences that would generate electricity.

So the Galvanists went back to the lab, devising ingenious demonstrations in which the conducting arc consisted of a glass vessel filled with unadulterated mercury. A dissected muscle was floated on the surface, with its spinal cord suspended from a silk thread. The thread was lowered so that the nerve touched the mercury, and—zap—the muscle twitched.

Impurities, Volta insisted. If the muscle moved there had to be dissimilarities in the metal—a circular argument that was impossible to refute.

They were at a standoff. For one man, the frog generated electricity that flowed through the metallic arc. For the other, the metallic arc generated electricity that flowed through the frog.

THE ONLY recourse for the Galvanists was to get the metal out of the loop. One experimenter showed that a piece of carbon served just as well: "Why then ascribe to the different power of metals, effects which can be produced by bodies which certainly have nothing of the metallic quality?" Volta insisted that the experiment proved nothing since carbon was, after all, a conductor.

Another experimenter showed that he could produce the galvanic response simply by touching the frog's muscle with one of his hands and the animal's severed nerve with the other. "Each time I touch it, the frog jerks, leaps, and, I'm tempted to say, escapes me." The conclusion seemed obvious: "metals are not the motors of electricity. . . . They possess no secret, magic virtue."

In what seemed the most persuasive experiment yet, Galvani eliminated external conductors entirely, gently manipulating a dissected frog so that the dangling sciatic nerve came directly into contact with the muscle controlling the leg. It immediately gave a kick. Where did the electricity come from but the animal itself?

A confident Galvani mocked Volta with his own words: "But if that is how things are—if such electricity is indeed wholly specific to the animal, and not common and extrinsic—what will become of the opinion of Signor Volta?"

It simply had to be modified. By now Volta was coming to think of the muscle, the nerve, the experimenter's hands, and even the frog itself as weak "second-class" conductors.

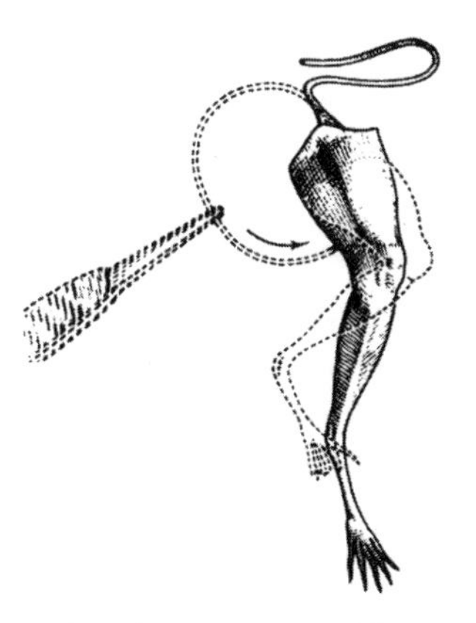

Galvani's experiment without
external conductors

Whether nerve was touched to muscle or silver to brass, the effect was the same: dissimilar conductors produced what he now was calling contact electricity.

In Galvani's earlier experiments a pair of first-class conductors—metal scalpels, brass hooks, silver box lids—were separated by a moist second-class conductor, the frog. He might just as well have used wet cardboard or, as Volta went on to show, a human tongue. Put a silver coin on top and a copper one on the bottom and you could taste electricity. The experiments involving a single metal were as readily explained. One first-class conductor formed an arc between two second-class conductors: the nerve and the muscle. Finally you could make an arc entirely from mushy second-class conductors: a hand and a frog. Organic or inorganic—it didn't matter, as long as the dissimilarity was there.

WE KNOW now that both men were right. They each proved it with a beautiful experiment.

First Volta. Taking several dozen disks, half made of copper and half of zinc, he stacked them one on top of another, alternating metals and separating them with round cardboard spacers that had been dipped in salty water. If he made the stack high enough he could give himself a mild shock. He could also use silver and tin, or replace the cardboard with little cups of salt water, chained together with bimetallic electrodes.

He had invented the battery. The title of his paper, published in 1800, seemed to say it all: *On the Electricity Excited by the Mere Contact of Conducting Substances of Different Kinds.* Galvani's frog was nothing but a moist separator in a voltaic pile.

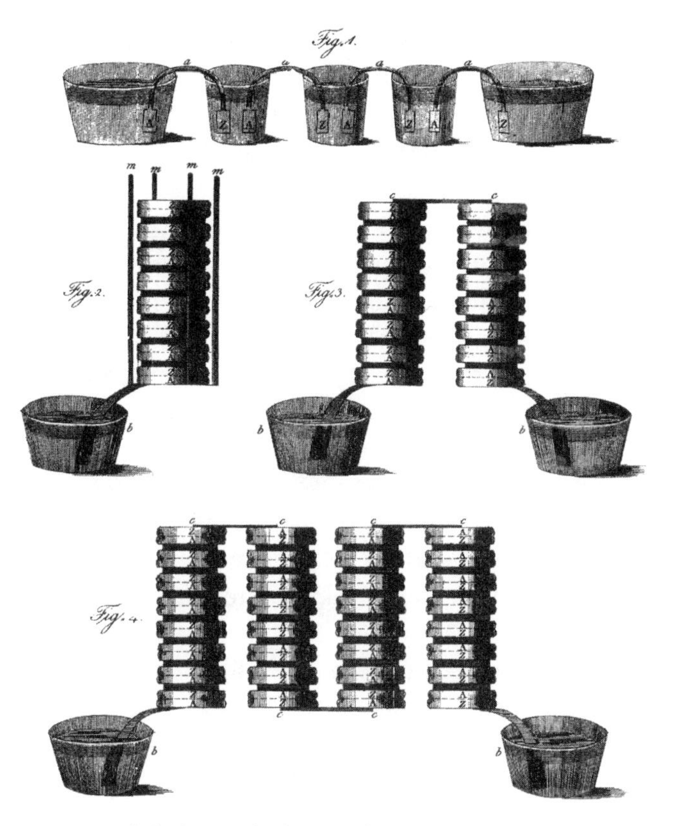

Volta's electrical pile. From his 1800 manuscript

But no, for Galvani's crowning experiment was as elegant as Volta's. He prepared another of his frogs "in the usual manner," with the primary nerve of each leg sticking out. In the earlier experiment he had touched nerve directly to muscle. This time, using a small glass rod, he nudged one nerve against the other: two identical conductors, and the result was a muscular contraction, one that did not occur if he simply irritated the second nerve with the piece of glass. "Now what dissimilarity could be called in to explain the contractions," he asked, "since the contact is formed between the nerves alone?" The effect could have been produced, he insisted, only "by a circuit of electricity inherent in the animal."

Though neither man could quite see it, their experiments complemented each other, for they were dancing around a single truth. Natural, artificial, animal—electricity is electricity. Volta didn't appreciate that what he was observing with his "contact electricity" was a chemical reaction (he actually thought his battery was a source of perpetual motion), and Galvani clung to the idea that there was something inherently different about biological electricity.

It would be years before physiologists laid out the details of what Galvani, egged on by Volta, had glimpsed with his frogs: how, in an organism, each microscopic cell acts like a little battery, with membranes behaving like cardboard spacers and charged ions playing the role of zinc and copper coins. What results is a standoff between positive and negative, the electromotive force called voltage. When a muscle moves or a finger feels the surface of a stone, a current flows through the nervous system. There is no ethereal "vital force." Life is electrochemistry.

CHAPTER 6

Michael Faraday

Something Deeply Hidden

Michael Faraday

I shall never look at the lightning flashes without recalling his delight in a beautiful storm. How he would stand at the window for hours watching the effects and enjoying the scene; while we knew his mind was full of lofty thoughts, sometimes of the great Creator, and sometimes of the laws by which He sees meet to govern the earth.

—Margaret Reid, Michael Faraday's niece

Spark.—The brilliant star of light produced by the discharge of a voltaic battery is known to all as the most beautiful light that man can produce by art.

—Michael Faraday, *Experimental Researches in Electricity*

Everybody knew Ada Lovelace was trouble. The daughter of the poet Byron, she had been born with a wild streak that her mother tried to suppress by occupying the girl's mind with mathematics. The therapy wasn't entirely successful—she tried to run off with one of her tutors. She was caught, tamed, and married to a nobleman, but she preferred the company of scientists. The inventor Charles Babbage was part of her coterie. He called her his "Enchantress of Numbers." She called herself the "Bride of Science." She was obsessed with new ideas: phrenology, mesmerism, a "calculus of the nervous system." In 1844, when she was twenty-eight, she struck up a flirtatious correspondence with England's greatest experimenter, Michael Faraday, proposing that she be his muse and "ladye-fairy."

> I will be the beautiful phantom, glowing in colour & eloquence, when you so order me. But I will now be a little quiet brown bird at your side, and gently let you teach me how to know & aid you. But my wand is yours at pleasure, & into your hands I deliver it for your use.

It is hard to tell from his careful replies what Faraday thought of her gushings, underscored with stabs of ink. He was fifty-three years old, married, a pious Christian, and in recovery from what would nowadays be called a nervous breakdown. Most of his great work was behind him—the experiments drawing together electricity and magnetism. Maybe it was Ada's flattery that pushed him to go one step further and show, in an elegant demonstration, that electromagnetism itself is intimately connected with light.

Lady Ada Lovelace

They had come from different worlds. The son of a blacksmith and a bookbinder's apprentice, Faraday had persuaded the great English chemist Humphry Davy to take him on as a secretary and assistant. The duties at first included serving as Davy's valet, traveling with him to Europe, and meeting the likes of Volta and André-Marie Ampère. Hired by the Royal Institution in London, Faraday embarked on a career doing the yeoman's work of science: analyzing clays for the Wedgwood china makers and gunpowder for the East India Company, studying industrial processes at metal foundries in Wales. When he was about the age of his young correspondent, an insurance company had asked him to report on the flammability of whale oil, and the British Admiralty on the best ways of drying meat. It was around that time, in late 1820, that Davy came to him with exciting news from a Danish scientist, Hans Christian Oersted.

Oersted had made a voltaic battery by filling twenty vessels with dilute acid and linking them in series with pieces of copper and zinc. Then he connected one pole of the apparatus to a long wire and placed it over a compass, parallel to the needle. The moment he touched the other end of the wire to the opposite side of the battery, the compass needle swung west. If he placed the wire beneath the compass, the needle swung east.

Overcoming their disbelief, Davy and Faraday rushed to repeat the demonstration, while Ampère, working in Paris, showed that parallel wires carrying currents in the same direction attracted each other like magnets. If one of the currents was reversed, the wires moved apart.

So clear a connection between magnetism and electricity was surprising enough. What was astonishing was that a force could move in circles instead of straight lines. ("Vertiginous electricity," one scientist called it.) Nothing in Newtonian mechanics had predicted this. Faraday went on to show

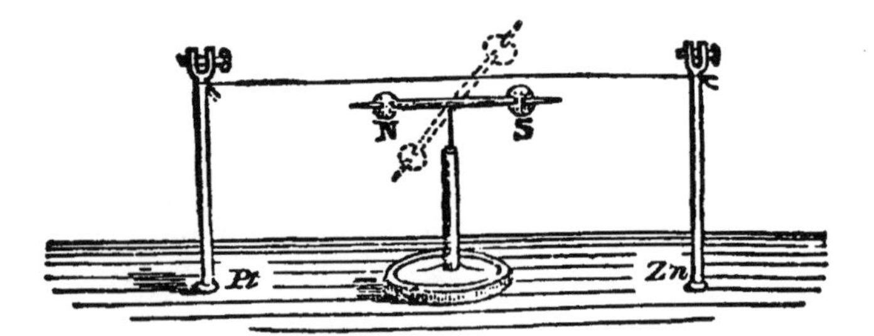

Oersted's experiment

that with a crude apparatus using mercury and a cork he could make an electrified wire revolve around a magnet or a magnet around an electrified wire. He had invented the electric motor. If he shaped a wire into a loop and attached it to a battery, it became a weak magnet. If he wound the wire into a spiral, the magnetic force was even stronger, concentrated inside the center of the coil.

With a few deft experiments, he had moved to the forefront of European science. And that is where he left things for a while. The next decade was dominated by the metallurgy of steel and copper, the manufacture of glass—more errands of the Industrial Age. In a letter to Ampère, he lamented how many of his days were "unfortunately occupied in very common place employment" instead of the research he loved. He found some time for more imaginative pursuits, studying the undulating patterns, or "crispations," that appeared when he spread a thin layer of sand or powder across the surface of a metal plate and vibrated its edge with a violin bow. A second plate of powder placed nearby would

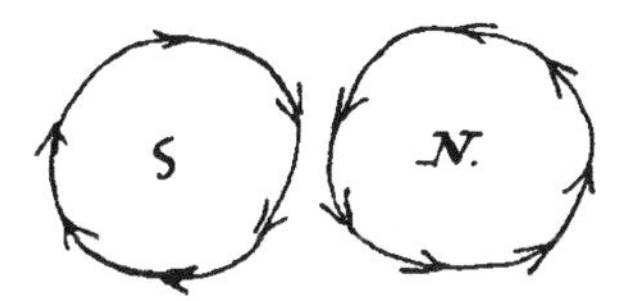

From Faraday's diary, a wire rotating around a magnet

vibrate in sympathy. He also experimented with liquids. "Mercury on tin plate being vibrated in sunshine gave very beautiful effects of reflection," he reported, sounding a bit like Newton. "Ink and water vibrated in sunshine looked extremely beautiful." It was not until 1831 that he finally returned to his coils and batteries.

By then the English electrician William Sturgeon had wound bare wire around a varnished iron core to make an electromagnet strong enough to hold more than its own weight. Using insulated wire, the American Joseph Henry made an electromagnet that would support more than a ton. One summer day Faraday decided to see what would happen if he put two separate coils in close proximity. He asked the shop at the Royal Institution to forge a ring-shaped iron frame seven-eighths of an inch thick and six inches in diameter. Around one side he wrapped seventy-two feet of copper wire, insulated with twine and calico. This he called coil A. On the other side of the ring, with about sixty feet of wire, he wound coil B.

There was no direct connection between one coil and the

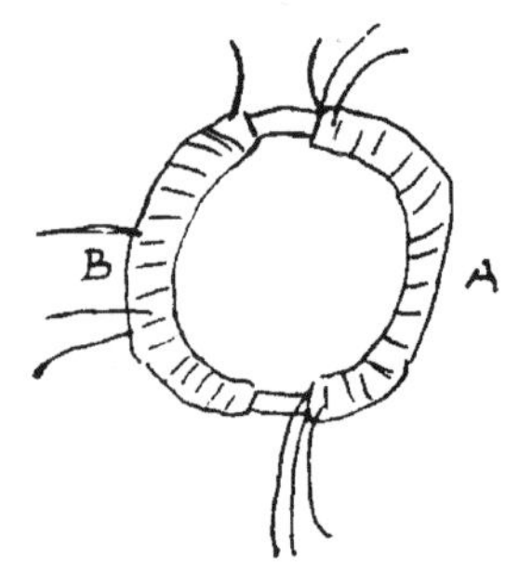

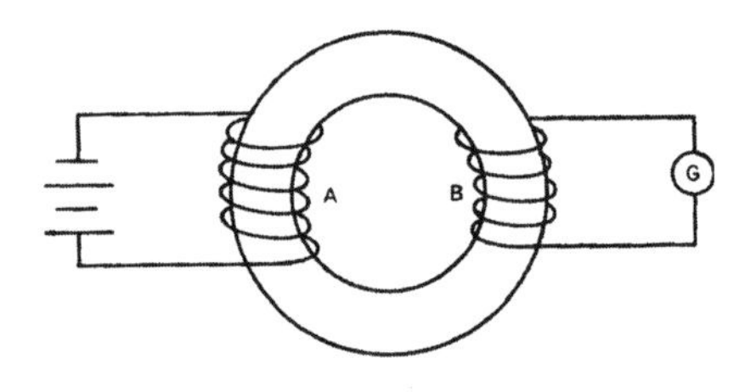

Faraday's drawings
of an induction ring

other. Yet when he touched the wires of the first coil to the poles of a battery, a galvanometer attached to the second coil twitched and oscillated before settling back to its original position. Disconnecting the battery made the needle move again. Thinking, perhaps, of the crispations in his acoustic experiments, he envisioned a "wave of electricity" produced in the primary coil traveling through the ring and somehow begetting a current in the second coil. He had discovered electromagnetic induction and cracked open a window onto a new world.

Moving a bar magnet back and forth inside a hollow coil also generated a current in the wire. Oersted had converted electricity into magnetism, and now Faraday had converted magnetism into electricity—producing the first crude electrical dynamo, the mechanical inverse of the motor he had devised ten years before. Electricity could be used to produce motion, and motion to produce electricity. Deep beneath the surface of reality, as Einstein would later say, something lay hidden. The job of the scientist was to coax it out.

The closer Faraday looked, the more he understood. He had noticed that over time, the copper electrodes in his voltaic cells slowly become tainted with zinc oxide, while the zinc electrodes became coated with copper. The flow of electricity from the battery's two poles must be accompanied by an internal movement of atoms. Not only was this the basis for a promising industrial process—copper-cladding or silverplating a piece of metal: the phenomenon also pointed toward another deep connection. A battery was a crucible for turning one kind of energy—chemical—into another kind: electrical. The process also worked the other way. When two electrified wires, positive and negative, were immersed in a slightly salty solution, hydrogen accumulated at one pole and oxygen at the other. Electricity was producing chemical reactions, and chemical reactions were producing electricity.

Scientists all over Europe were confronting these mysteries. Was water made from hydrogen and oxygen? Or, as a German scientist proposed, was water elemental—with oxygen made from combining it with positive electricity and hydrogen from combining it with negative electricity? He even tried to revive the phlogiston theory. It was Faraday, more than anyone, who cut through the confusion. Throughout the 1830s he demonstrated in one experiment after another how electricity, magnetism, and chemistry all were related. Then, a few years before Ada Lovelace began her pursuit, he fell into a slump.

For a long time he had complained of problems with his memory. Now he was receding into a dark depression, unable to concentrate, suffering from dizzy spells. Maybe the cause was mental fatigue—or cumulative poisoning from all the chemicals that had touched his skin. On doctor's orders he began turning down speaking invitations and requests for industrial research, confining himself mostly to writing and contemplation. A falling-out with his church—apparently over some kind of factional dispute—added to his isolation. Then came the barrage of italicized flattery from Lady Lovelace, tempting him so strongly that he felt he had no recourse but to cut it off: "You drive me to desperation by your invitations," he pleaded. "I dare not and must not come and yet find it almost impossible to refuse."

Maybe it is too great a reach to say that his close encounter with the Bride of Science was a turning point, but it was around this time that the clouds began to lift. Faraday, a burnt-out case, returned to his laboratory to take up a question that had been gnawing at him for years. It was clear now that electricity and magnetism were tightly related. But could there also be a connection between electricity and light?

As scientific adviser to Trinity House, an organization chartered in 1514 by Henry VIII "so that they might regulate the pilotage of ships in the King's streams," Faraday had worked to improve the powerful Argand oil lamps used in the lighthouses along the English and Welsh coasts. In late August 1845, he fired up one of the beacons in his laboratory and prepared the way for what would become his most beautiful experiment.

As light travels it vibrates transversely—at right angles to its direction of motion. But if it is reflected from a flat surface or passed through certain crystals like tourmaline, it becomes polarized, its oscillations confined to a single plane.

If you looked at one of these beams through a second polarizing crystal while rotating it through 360 degrees, the image would go from light to dark to light again as the filters moved in and out of sync.

The question Faraday now posed was whether an electrical current could twist a light beam, making its plane of vibration rotate. Filling a long trough with a mildly conducting solution, he placed platinum electrodes at each end and connected them to a five-cell battery. The setup was similar to what one might use to decompose water into its constituent gases or to copperplate a spoon. He lit the Argand lamp and reflected its light off a glass plate, causing it to become polarized. Then he passed the beam through the same solution where electricity was flowing and rechecked the polarization using a device called a Nicol prism.

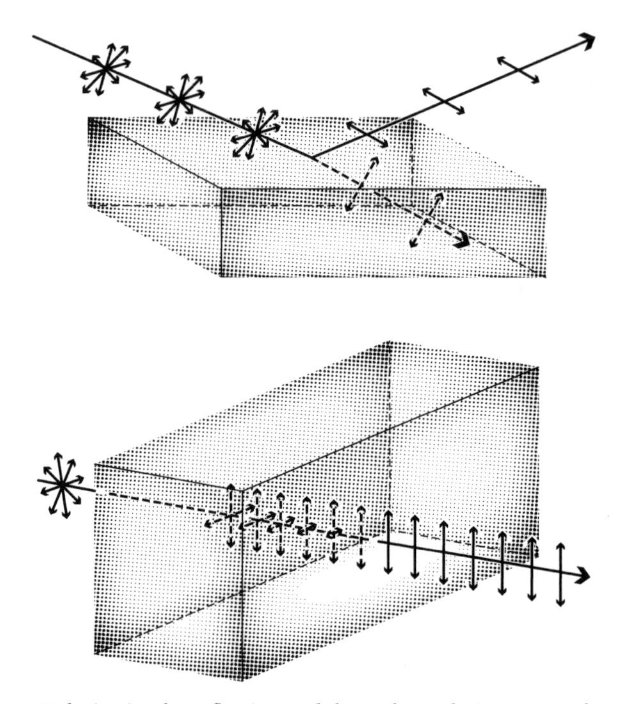

Polarization by reflection and through a polarizing crystal

Nothing happened: the direction of vibration was unchanged. He tried the experiment with continuous currents, with intermittent currents, with currents passed through various solutions, but there was no perceivable effect. He tried shining the light beam parallel to the electrical flow instead of across it. Still no shift in polarization. Speculating that his batteries were not strong enough, he tried again with a static electricity generator, charging a plate of glass and shining the light beam through it every which way. Still nothing.

It was then that he decided to try magnetism. Retrieving

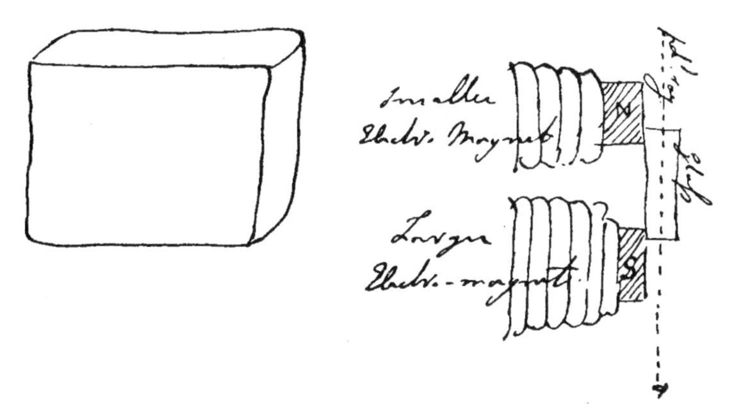

The polarization experiment. A square of glass (left) is placed against the opposing poles, north and south, of an electromagnet. A polarized light beam passing through the glass is rotated by the electromagnetic field. From Faraday's diary

from his stockpile a heavy piece of optical glass about two inches square and half an inch thick, he mounted it next to the poles of a powerful electromagnet. He arranged the lamp and polarizing surface so that horizontal light waves passed through the length of the glass. Looking through the Nicol prism, he rotated it until the beam was extinguished. Then he switched on the current. The image of the flame suddenly reappeared. He turned the magnet off and the flame disappeared again. The magnetic field was making the light beam turn.

All his previous work with magnetism and electricity was coming to a climax. With the exhilaration that comes from total absorption, he threw himself into his research. "At present I have scarcely a moment to spare for any thing but work," he wrote to a colleague. "I happen to have discovered a direct relation between magnetism & light also Electricity &

light—and the field it opens is so large & I think rich that I naturally wish to look at it first. . . . I actually have no time to tell you what the thing is—for I now see no one & do nothing but just work."

The alignment of the magnetic field, Faraday learned, was paramount. Nothing happened when he placed a north magnetic pole on one side of the glass and a south pole on the other. Nor was there any effect when he exposed the two sides of the glass to similar poles, or when the same poles were lined up on the same side. "BUT," he wrote in his diary (like Ada Lovelace on adrenaline, he underlined the word three times), "when contrary magnetic poles were on the same side, there was an effect produced on the polarized ray, and thus magnetic force and light were proved to have relation to each other."

He confirmed that a powerful permanent magnet would also rotate the beam and that other transparent materials could be used in place of glass. Some worked better than others, but in every case the degree of rotation depended on the strength of the magnetic field. And if the polarity of the field was reversed, the light beam turned the other way. The final thread had been pulled into the tapestry. Electricity was entwined with magnetism and magnetism with light.

It was left for James Clerk Maxwell, two decades later, to show with his famous equations that light *is* electromagnetism. With barely a pause Faraday tried to push the unification further, attempting to connect gravity with magnetism, a quest that eluded him and Einstein and every scientist since. "ALL THIS IS A DREAM," he wrote in his diary. "Nothing is too wonderful to be true, if it be consistent with the laws of nature, and in such things as these, experiment is the best test of such consistency."

Throughout all this, Ada was still in his mind. "You see what you do—ever as you like with me," he wrote to her in 1851, six years after he had begged her to go away. "You say write & I write—and I wish I had the strength and had rest enough for a great deal more." The next year she died of cervical cancer. She was thirty-six. Faraday outlived her by fifteen years.

CHAPTER 7

James Joule

How the World Works

James Prescott Joule

You will therefore be surprised to hear that until very recently the universal opinion has been that living force could be absolutely and irrevocably destroyed at any one's option. Thus, when a weight falls to the ground, it has been generally supposed that its living force is absolutely annihilated, and that the labour which may have been expended in raising it to the elevation from which it fell has been entirely thrown away and wasted, without the production of any permanent effect whatever.

—James Joule, lecture in Manchester, 1847

Joule

WE DON'T know what William Thomson was thinking on that cloudy August day in 1847 as he set out on foot from Chamonix toward Saint Gervais, but it probably had something to do with physics. A child prodigy, he had published his first scientific paper when he was sixteen. Fresh out of Cambridge at twenty-two, he was named to the chair of natural philosophy at the University of Glasgow, and now, a year later, was trekking in the French Alps bound for Mont Blanc. All the forces of nature, Thomson was coming to believe, must be related (he had been "inoculated with Faraday fire"), and he may have been turning that thought over in his head as he approached the turnoff for the high trail over the Col du Bonhomme and encountered the familiar face of another hiker, James Prescott Joule.

Joule was on his honeymoon (his wife was following behind in a carriage), and he was carrying, or so Thomson would later remember, a long thermometer with which to measure the temperature of waterfalls. If Joule was right, the water at the bottom of a fall must be slightly warmer than the water at the top, and this would mean that the prevailing theory of heat, the one force of nature Thomson found most puzzling, was incorrect. He agreed to meet up with Joule a few days later at the "Cascade de Sallanches"—probably the 1,199-foot Arpenaz waterfall, which by Joule's reckoning should show a temperature difference of roughly one and a half degrees Fahrenheit. There was too much spray to take an accurate reading, Thomson reported. Dataless, the two men went their separate ways.

The story is probably a little too neat. Though he did meet Joule on the trail, Thomson, the future Lord Kelvin, made no mention of the thermometer when he wrote to his father a few days later from the Hospice du Grand-Saint-Bernard. Memories have a way of becoming mushed together. It seems likely that years later, when Kelvin, by then one of the most revered scientists in Europe, described the encounter, he was conflating it with an earlier event.

Their paths had first crossed two months earlier at a scientific meeting in Oxford. Accustomed to having his ideas ignored, Joule, a self-taught amateur from the industrial city of Manchester, was delighted when at the end of his talk this young man named Thomson stood up and made some acute observations. Joule was too awkward and reserved to make a good lecturer, but at least someone had been listening. Thomson would later insist that he had remained seated and asked his questions only afterward. This time maybe Joule's memory was playing tricks, but the experiment he described had clearly made an impression.

Lavoisier had loosened the grip of the fictional phlogiston, but before his death he introduced another invention: caloric, his name for an invisible substance—a "subtle fluid"—said to be the carrier of heat. The idea seemed sensible enough. Something that was hot was dense with caloric, and because caloric had a tendency to expand, it would naturally migrate toward where it was not. Put a metal poker into a fire, and the caloric will rise up the shaft until you can feel the warmth in the handle. Things expanded when they were heated because they took in caloric. Gases got hotter when compressed because the caloric within them became more concentrated, and they cooled as they rarefied because the caloric spread out.

In a steam engine, caloric could even be harnessed, like

water in a mill, to do work. Concentrated in lumps of burning coal, caloric flowed into the boiler, heated the water, and was carried with the steam that pushed the piston. When the cycle was complete the same amount was expelled into the air as exhaust. Like matter, caloric could be neither created nor destroyed. The universe had been bequeathed with a fixed amount that was constantly being shuttled from place to place.

That is why Thomson had found Joule's presentation so unsettling. Joule claimed to show that heat could be created at will. At a reception later that day at the Radcliffe Camera, the elegantly domed cylindrical annex of the Bodleian Library, they discussed the implications. "Joule is, I am sure, wrong in many of his ideas," Thomson wrote to his father a couple of days later, "but he seems to have discovered some facts of extreme importance." Not long afterward, Joule followed up with a letter to his new friend suggesting how a rope, a bucket, and a good thermometer could be used to show that heat was generated even by falling water.

JOULE was not the first scientist to challenge the notion that heat is an invisible fluid, and here Lavoisier, or rather his widow, Marie Anne, enters our story one last time. She too had spent time in prison, but after Robespierre's fall had reclaimed the Lavoisier estate and was presiding again over a lavish salon frequented by some of Europe's premier thinkers. One of her guests was Benjamin Thompson, an American exile who had found himself on the losing side of the Revolution and had fled to London, abandoning his wife and daughter. He later moved to Bavaria where he acquired a title, Count Rumford, and after he met Marie Anne, in 1801, was determined to acquire her as well. She was lively, kind, and intelligent, he wrote, and though she was, as he delicately put it, rather "en bon point" (pleasingly plump), "her personal fortune is considerable."

Arrogant and moody, the count was no prize himself (his previous bride had also been a rich widow), and he must have sensed that the way to Marie Anne's heart was through her brain. He courted her with tales of his scientific feats, many of which had to do with heat—the invention of the Rumford stove, thermal underwear, a drip coffeepot, and most significantly the first widely known experiment to cast doubt on the caloric theory.

While working with the Bavarian military, Rumford had been impressed by how much heat was produced by boring out the holes of brass cannons. Conventional wisdom held that the drilling was liberating caloric that was trapped in the metal, but Rumford was dubious. He submerged a cannon in water and harnessed two horses to turn the bit. The water got hotter and hotter until after two and a half hours it came to a boil "merely by the strength of a horse, without either fire, light, combustion, or chemical decomposition."

"It would be difficult to describe the surprise and astonishment expressed in the countenances of the by-standers, on seeing so large a quantity of cold water heated, and actually made to boil without any fire," he reported to the Royal Society. He saw no reason to doubt that, as long as the horses lasted, he could keep on churning out more heat. If there was such a thing as caloric, the cannon itself seemed to hold an inexhaustible supply.

Others had come to a similar conclusion: that heat is not a

material thing but some kind of *vis viva* ("living force") or motion—"a very *brisk* and *vehement agitation* of the parts of a body," Robert Hooke had written. The Swiss mathematician Daniel Bernoulli had proposed that heat was the vibration of invisibly tiny particles of matter. But that was a dying theory, and Rumford's experiment hadn't been done with enough precision to change many minds.

After a four-year courtship, Rumford persuaded Marie Anne to be his wife and moved into her mansion. The marriage didn't last. One day, jealous of his solitude, he barred her guests from the house. She retaliated by pouring boiling water, rich in caloric, on his roses. Finally she paid him 300,000 to 400,000 francs to go away.

DURING THE FIRST decades of the nineteenth century, as experimenters like Faraday teased out hidden electromagnetic connections, the nature of heat—so familiar, mundane, and powerful—remained stubbornly obscure. Somehow in its passage through a steam engine, this mysterious nothing could literally move the earth. Steam-driven pumps sucked tons of water from mine shafts, exposing deep veins of coal that would drive locomotives, factories, and mills. Steam shovels excavated lodes of iron ore from which to forge more tools and machinery. With a source of power so abundant and portable, a small water-driven industrial economy that had sprouted along northern England's millstreams began spreading southward into the flatlands. In Manchester, where Joule was born in 1818, there were soon steam engines everywhere, belching smoke and turning wheels. The basic principle of these devices was well

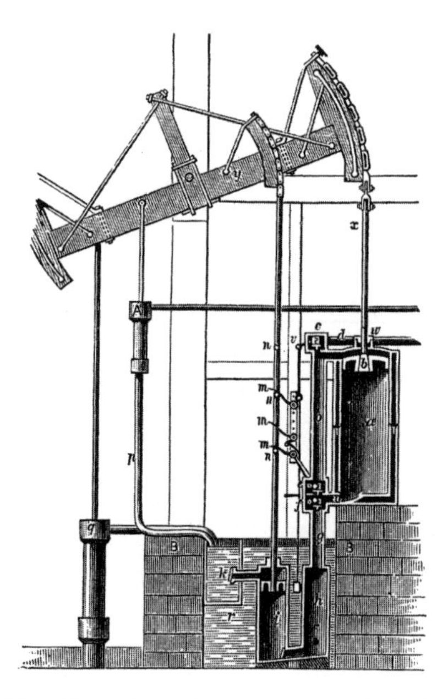

A late-eighteenth-century steam engine made by James Watt

understood—a head of high-pressure steam pushed a piston that was geared to turn a wheel—but no one knew what laws of nature made this possible. It was as though, later on, the nuclear reactor had been developed through trial and error without anyone understanding the physics.

It seemed clear enough what was happening in the old riverside mills. Water flowed rapidly at the top of a paddle wheel, fell downward, and emerged at the bottom at a slower pace. Some of its "effort," or *vis viva,* was spent turning the wheel. The greater the difference between the incoming and outgoing velocities, the more power was extracted from the waterfall.

A water wheel

Engineers like the Frenchman Lazare Carnot had studied how to make water mills as efficient as possible. In 1824 his son, Sadi Carnot, named for a Persian poet, went on to propose an analogy: a steam engine is like a paddle wheel with the water replaced by caloric "falling" down a gradient from hot to cold. He described his theory in a treatise, little known at the time, called *Reflections on the Motive Power of Fire.* Steam entered the engine at a very high temperature and exited at a much lower one. By maximizing the difference one could squeeze as much work from the fuel as physics allowed. One could also run the cycle backward: performing work to pump heat back uphill (what a modern refrigerator does with the power it sucks from the wall outlet).

Carnot's analysis marked the beginning of what Kelvin

would name thermodynamics, but it left intact the idea that heat was a substance—caloric—that, like water passing through a water wheel, was neither created nor destroyed. As a teenager, Joule probably learned all this from his tutor, John Dalton, another Manchester resident, whose chemistry experiments had established the foundation of modern atomic theory. Joule's father, a prosperous brewer, had arranged for James and his brother to study privately with the chemist. Quickly becoming the eager boy scientist, James shocked his playmates with Leyden jars and experimented with electricity on a lame horse and on a servant girl, who received such a jolt that she fainted. By the time he was nineteen, working at the brewery, he was tinkering with coils and magnets, hoping to invent an electric motor more powerful than a steam engine yet cheaper to run.

To power the device, Joule used voltaic cells in which two electrodes—one zinc and one copper—were immersed in dilute sulfuric acid. In a battery like this, the acid eats away at the zinc releasing an excess of electrons. Connect a motor across the poles and a current will flow, magnetizing the coils that make the rotor spin.

Early on, Joule noticed that the strength of an electromagnet increased as the square of the current. By doubling the number of batteries, you quadrupled the power. The possibility that the same might be true for an electric motor must have seemed as stunning as cold fusion did in the 1980s. "I can hardly doubt that electro-magnetism will ultimately be substituted for steam to propel machinery," Joule declared with the enthusiasm of a twenty-year-old unused to the troubles posed by the material world. "The cost of working the engine may be reduced *ad infinitum.*" Except for minor impediments like air resistance and friction, he believed, "there seemed to be nothing to prevent an enormous velocity of rotation, and consequently an enormous power."

Reality was not so compliant. Joule's first motor was barely powerful enough to turn itself over. He tried different arrangements of coils and batteries, and wrapped different kinds of wire around different kinds of cores, but he continued to run up against nature's will. The more current you fed to the motor, the hotter its coils became. In fact Joule discovered that the heat also increased according to the rule of squares. If you doubled the number of batteries you quadrupled the heat. It was a losing proposition. The hard truth was that you cannot get more energy out of a system than you put into it. You can only convert the energy into a different form.

By 1841, the lesson had thoroughly sunk in. The best steam engines in the world could sap enough *vis viva* from a pound of coal to lift 1.5 million pounds one foot off the ground—or one pound 1.5 million feet off the ground. A pound of coal, in other words, was doing 1.5 million foot-pounds of work. Joule's best battery-powered motor could extract only one-fifth as much from a pound of zinc, and zinc cost sixty to

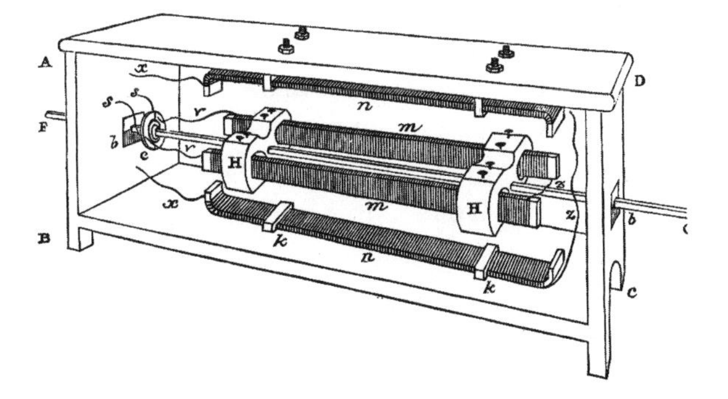

Joule's electric motor. From his *Scientific Papers*

seventy times more than coal. "The comparison is so very unfavourable," he lamented, "that I confess I almost despair of the success of electro-magnetic attractions as an economical source of power."

Today, of course, electric motors, powered by electricity from the grid, have supplanted steam engines in factories around the world. But ultimately their energy comes from steam. In a power plant, coal or gas is burned or uranium is fissioned to boil water, moving the turbines that drive the dynamos that make electricity.

To a person engaged in the practical craft of motor making, heat was a nuisance, but Joule was starting to sense a deeper truth: that there was a fundamental connection between heat and work. A wire shorted across the poles of a battery will quickly become so hot that the insulation smokes. But if you insert a motor into the circuit, the wire stays cooler: work is accomplished at the expense of heat. The same was true if you used the battery to electrolyze water, splitting it into hydrogen and oxygen, or to electroplate a spoon.

Maybe caloric was flowing from the battery, along with electricity, but the battery didn't seem to get cooler—more evidence that the heat was not preexisting but generated on the fly. In 1843 Joule began putting the hypothesis to a test.

The idea was to place a coil inside an insulated glass tube filled with water and spin it with a hand crank. Sitting alongside would be two powerful electromagnets salvaged from Joule's electrical engines. The result was a generator. The wires of the coil were connected to a galvanometer to measure how much current was produced. (To keep the wires

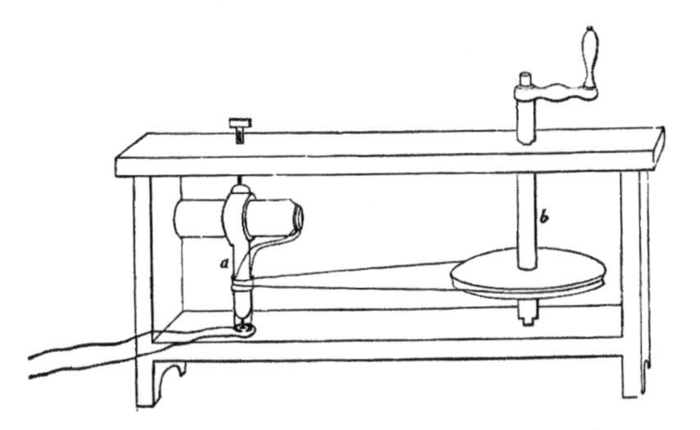

Joule's generator. The electromagnets are not shown.

from twisting he devised a clutch made from mercury sitting inside two semicircular grooves.) He would measure the temperature of the water, steadily turn the crank for fifteen minutes, and then take the temperature again.

It was a very delicate operation. He had to adjust for things like the cooling effect of the air and changes in room temperature. He had to allow for the fact that the current induced in the spinning coil was not steady but pulsating. He tried different strengths of magnets, different numbers of batteries, and when he was done he had persuaded himself that the spinning made the water slightly warmer. Comparing the readings from the galvanometer with those from the thermometer, he saw a familiar relationship: double the current and you get four times the heat.

The coil was not connected to a battery. So where was the caloric coming from? The only possible source of heat was the work Joule was doing by turning the crank. As in Rumford's cannon experiment, circular motion was being converted into a different kind of movement—tiny material vibrations our fingers feel as heat.

To persuade the skeptics Joule knew he would have to go a

step further. Precisely how many foot-pounds of work does it take to produce a given amount of heat? He redesigned his original apparatus, winding the axle of the hand crank with two long pieces of twine, coiling them in opposite directions. Each was slung over a pulley and attached to a pan that held a weight. As the weights fell, the coil would spin and generate electricity and heat.

After trying different weights falling from different heights (to give them enough room he dug two holes in his garden), he estimated that the mechanical effort stored in an 838-pound mass suspended a foot off the ground would produce enough heat to raise a pound of water by one degree Fahrenheit. Pound for pound, the temperature of a waterfall 838 feet high—King Edward VIII Falls in Guyana comes close—should be roughly one degree warmer at the bottom than at the top.

In August 1843, he described his results at a scientific conference in Cork, Ireland, but, as he later put it, "the subject did not excite much general attention." The tangle of different phenomena—electricity, magnetism, heat, motion—may have obscured the point of his presentation, and Joule

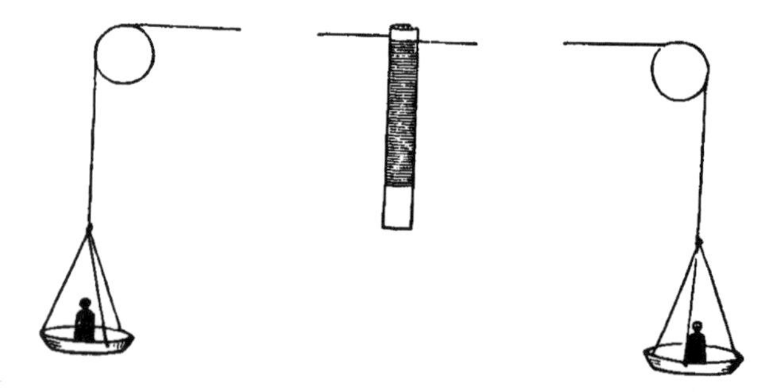

Weights and pulleys to turn the generator crank

himself probably didn't help. He still needed a knockdown experiment that would speak for itself—one that was simpler, more elegant, with cleaner lines.

BY THE TIME of the Oxford meeting in 1847, where he met William Thomson, Joule had his proof in hand. It was late in the afternoon so he was asked to keep the presentation short. He had carried his new apparatus down from Manchester and set it up on a table in the lecture room: a vessel made of copper clad with tin. The lid, also tin, had a hole cut dead center to accommodate the shaft of a brass paddle wheel, and another hole in which to insert a thermometer.

Joule explained how he had filled the vessel with water and rigged up the weights, strings, and pulleys to make the paddle turn. Around the inside wall of the container, brass baffles resisted the circular movement of the water, increasing the friction. Placing a 29-pound weight in each pan, he raised them 5.25 feet from the ground and let them fall. Then he rewound the spindle and let the weights fall again, repeating the procedure twenty times. All together, the work used to churn the water added up to about 6,090 foot-pounds: 58 pounds of weight raised 105 (20×5.25) feet high. He con-

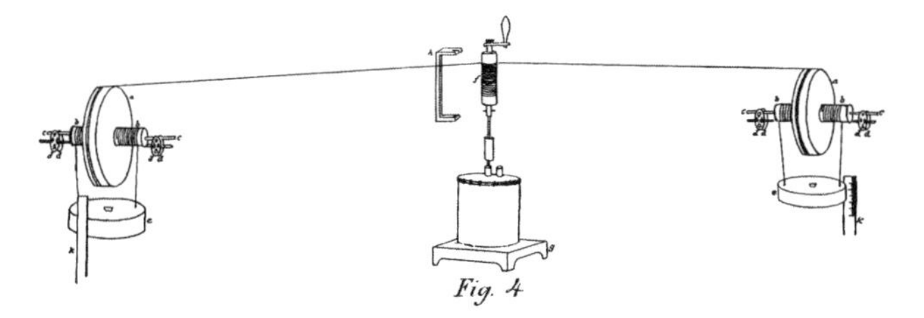

The refined version of Joule's experiment

ducted the experiment a total of nine times, finding in the end that the temperature of the water had risen by an average of 0.668 degree.

He figured that some of the force from the falling weights had been wasted overcoming the friction of the pulleys and string. To estimate how much, he took a roller of the same diameter as the spindle and wound a piece of the twine once around it, suspending his weights from both ends. Gradually adding smaller weights to one side, he found that it took about 7.2 ounces (3,150 grains) to upset the balance and cause the wheel to budge.

Taking this and other factors into account, he improved on his earlier calculation: heating one pound of water by one degree took 781.5 foot-pounds of effort, a figure he would later refine to 772 foot-pounds. Conversely a one-degree difference in temperature had the potential—if only you could tap it—of raising a one-pound weight 772 feet in the air.

This time there were no coils and batteries to muddy the message. Heat and work were not only related but the same: two different ways in which "effort" or *vis viva*—energy, we now say—was converted into motion. Work was what resulted when a force was used to move something across a distance—a horse pulling a wagon. It was structured energy put to productive use. Heat, on the other hand, was unproductive work, directionless, unstructured, energy dissipated as random microscopic vibrations. As atomic theory continued to develop, the image would become more vivid: heat is the vibration of atoms.

It was an extraordinary notion, just barely understood: Joule was expending this stuff called energy when he raised a weight off the ground, and when the weight fell it was giving the energy back. Harnessed to a generator, the work could be converted into electrical power, which might be used to run a motor and pump water uphill to a reservoir, where it could flow downward and turn a water wheel, which might be used to wind a giant clock spring. But at every step of the way, a portion of the energy would be lost as heat. And if the weight was simply allowed to fall with no work done, heat is all you would get—from the impact with the ground and the resistance of the air. It was not caloric that must always be conserved but energy.

Once he accepted Joule's discovery, Thomson went on to work out the implications. Though heat didn't disappear from the universe, it gradually became degraded, flowing from hot to cold and never back again—"irrecoverably lost."

The implication, he realized, was that the world had once been extremely hot and would inevitably become colder: "Within a finite period of time past the earth must have been, and within a finite period of time to come the earth must again be, unfit for the habitation of man."

The same was true for the universe. It began with a bang and it has been downhill ever since. All that from trying to understand steam engines.

CHAPTER 8

A. A. Michelson

Lost in Space

Albert A. Michelson

There are no landmarks in space; one portion of space is exactly like every other portion, so that we cannot tell where we are. We are, as it were, on an unruffled sea, without stars, compass, soundings, wind, or tide, and we cannot tell in what direction we are going. We have no log which we can cast out to take a dead reckoning by; we may compute our rate of motion with respect to the neighboring bodies, but we do not know how these bodies may be moving in space.

—James Clerk Maxwell, *Matter and Motion*

FOR AN old sailor like Albert Abraham Michelson, what Maxwell was describing was a nightmare—to be adrift on a windless night without a star to guide you. Michelson had learned his physics as a young man in the U.S. Navy, both at the Academy in Annapolis and on the ocean, practicing the art of navigation. You had to forget Copernicus, think like Ptolemy. You and your ship were at the center with the orbiting stars as your guide. In reckoning your position, you would take into account the velocity of your vessel, adjusting for the speed and direction of the wind. But as lost and confused as a young ensign might feel, he knew that his ship was in the crosshairs of some godly eye—precisely at a certain latitude and longitude. Surely the same must be true as we sailed the universe. There had to be some kind of standard, something fixed to measure by.

Or so he hoped. It was 1885 and for the past several weeks, Michelson himself had been unmoored, living at the Hotel Normandie in New York City under the care of a prominent psychiatrist. He'd gone soft in the head, as his collaborator Edward Morley put it—driven one moment, depressed the next. His wife tried to commit him to an asylum. His children were scared of him. The doctor ultimately decided that there was nothing dangerously wrong. But Michelson was clearly a man obsessed—by light and by color, by the way colliding beams caused the iridescent shimmer of an insect's wings. He imagined luminescent music, where the performer would sit at a keyboard and play visual notes from the spectrum, color chords and arpeggios, "rendering all the fancies, moods, and emotions of the human mind."

In November 1885, Michelson, in a manic mode, prepared to return to his laboratory at the Case School of Applied Science in Cleveland, only to find that his position had been filled and that he would have to take a salary cut. He came home anyway, moved into the back room of a house where he no longer felt wanted, and prepared for his greatest experiment—using light beams to clock the velocity of the Earth against the backdrop of outer space.

IN *Two New Sciences,* Galileo had suggested how one might test whether light is instantaneous or moves with a finite speed. Standing on a hilltop at night, an experimenter would flash a bright light toward a distant hill, where an assistant, awaiting the signal, would answer by flashing back. If there was no noticeable delay, one could conclude that "if not instantaneous, light is very swift."

No hills on earth are far enough to really tell, but in the 1670s the Danish astronomer Ole Roemer found a way to make the measurement across the solar system. Training his telescope on Jupiter at certain times of the year, he noticed that its innermost moon, Io, seemed to be slowing in its orbit. That, Roemer surmised, was because as Jupiter and its moons moved farther from Earth, their light took longer to reach us. Taking into account what was known about planetary distances, his observations implied a light speed of about 225,000 kilometers (140,000 miles) per second.

It was a bold conclusion—Kepler and Descartes had been sure that light moved infinitely fast—that was not confirmed until half a century later when an English astronomer, James Bradley, discovered a phenomenon called the aberration of starlight. Tracking the star Gamma Draconis, he found that it

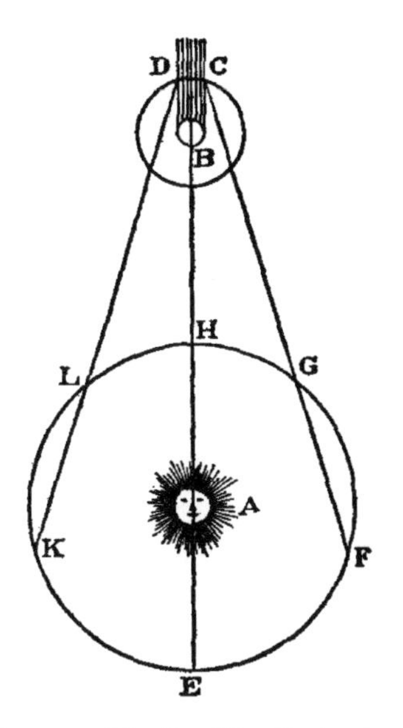

A diagram by Roemer of Jupiter (B) eclipsing its moon Io (DC) as viewed from different points in the Earth's orbit around the sun

wandered from its expected position, moving steadily southward from September to March and then northward again. After ruling out other possibilities, he hit on the explanation: by the time the starlight reached his telescope, the Earth had shifted position. Like a duck hunter leading with his rifle, an astronomer had to lead with his telescope. Based on Bradley's data, light traveled at 183,000 miles per second.

In 1849 the French physicist Armand-Hippolyte-Louis Fizeau made a more direct measurement with a sophisticated version of Galileo's flashing lanterns. From a house in the western suburbs of Paris he projected a light beam toward a mirror atop Montmartre, which reflected it back again. Interposed in the path was a rapidly spinning cogwheel with 720 precisely machined teeth. When the rota-

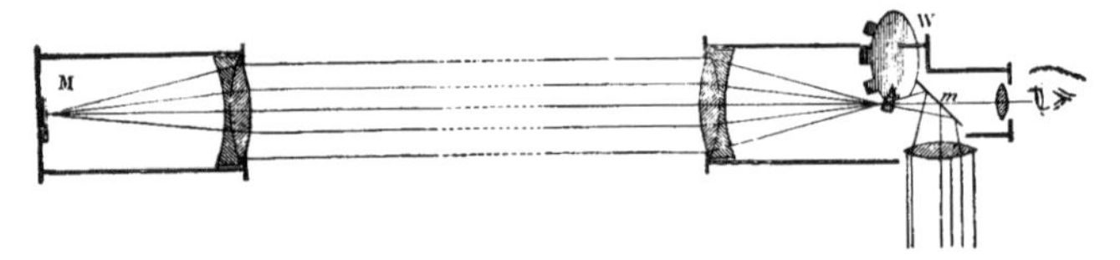

The Fizeau experiment. Light is projected between the teeth of a rapidly spinning cogwheel onto a mirror (M), which sends it back through the wheel again.

tional speed was set just so, the light, going and coming, would pass through a gap in the wheel's circumference and appear in Fizeau's eyepiece as "a luminous point like a star." Spin the wheel a little faster or slower and the beam would be eclipsed. From the length of the light path and the speed of the wheel, Fizeau estimated the velocity of light at about 196,000 miles (315,400 kilometers) per second.

Thirteen years later, his rival Léon Foucault refined the experiment, replacing the cogwheel with a spinning mirror set at an angle. On the two legs of its journey, the ray would strike the mirror at slightly different points in its rotation. Measuring the tiny displacement gave light speed as 185,000 miles (297,700 kilometers) per second.

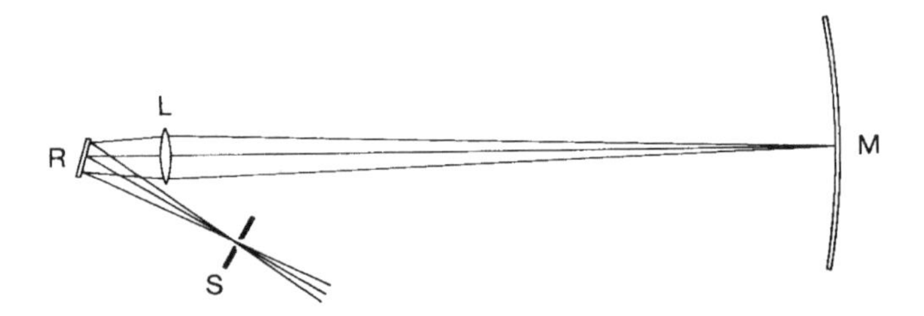

The Foucault experiment. Light from the source (S) glances off the spinning mirror (R), then travels through the lens (L) to a second mirror (M). By the time the beam returns, the first mirror has moved, causing a slight deflection.

Michelson would have learned all this at the Naval Academy in Annapolis, where he had arrived in 1869 by his own circuitous route. The oldest son of Polish immigrants, he had moved with his family to California where his father opened a dry goods store at a gold mining camp. Later they followed the silver rush to Nevada, and after high school Albert applied to the Academy. When he failed to get an appointment from his congressman, he had the temerity to catch a train to Washington and persuade President Ulysses S. Grant to intervene. By 1874 Michelson was an ensign aboard the USS *Worcester,* going on to become an instructor in physics and chemistry at Annapolis. It was there that he met Margaret Heminway, the niece of an officer who headed the physics department and the daughter of a Wall Street tycoon. They married in 1877, and a year later, with $2,000 from his father-in-law, Michelson began planning his first big experiment.

In Foucault's attempt to clock a light beam, the displacement from the spinning mirror was less than a single millimeter—very difficult to gauge. Michelson knew that if he could project the beam down a much longer path (Fou-

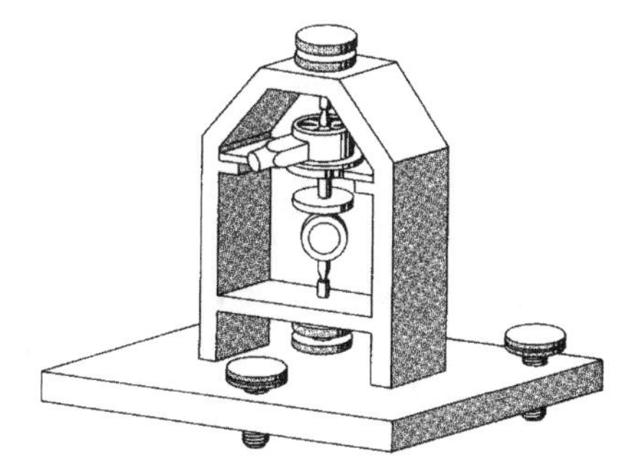

Michelson's drawing of his rotating mirror

cault's was just twenty meters long), the lag time would be that much greater. The returning beam would hit the mirror later in its cycle, resulting in a larger deflection and, he hoped, a better value for the speed of light.

He began by placing two mirrors, one revolving and one stationary, about 2,000 feet apart along the north seawall of the campus. To measure the separation precisely, he used a steel tape, calibrated against a copy of the "standard yard." Holding the tape flat along the pier with lead weights, and taking pains to ensure that it was stretched at a constant tension, he made several readings. Correcting for the effect of temperature on expansion and contraction of the tape, the distance between the mirrors came out to be 1,986.23 feet.

Everything had to be just so. To adjust the position of the stationary mirror, the one that would bounce the light beam back down the long course, he used a telescope and a surveying device called a theodolite. To clock the speed of the revolving mirror he used an electric tuning fork (which he had meticulously calibrated against a standard tuning fork). A small steel mirror was attached to one of the tines, reflecting an image of the spinning apparatus. When the frequency of vibration coincided with the speed of rotation, the image would freeze stroboscopically.

Using a steam-powered blower to spin the mirror at 256 revolutions per second and sunlight focused through a lens, he measured the deflection at the end of the light's journey at 133 millimeters—"being about 200 times that obtained by Foucault." A few calculations yielded a speed of 299,940 kilometers or 186,380 miles per second—just slightly higher than today's accepted value of 186,282.397. (So confident are scientists of that number that the meter is now defined in terms of the speed of light rather than vice versa.)

"It would seem that the scientific world of America is destined to be adorned with a new and brilliant name," the *New York Times* observed, predicting that light would soon be measured "with almost as much accuracy as the velocity of an ordinary projectile."

BY THE time Michelson was making his mark with his optical speedometer, scientists thought they had settled the question of whether light was made of particles or waves. Newton had imagined light as "globular bodyes" and even tried to explain refraction that way. Passing through a prism and reentering the air, different-colored particles would be given different spins, like "a Tennis-ball struck with an oblique Racket."

Harder to fathom was the phenomenon that came to be called Newton's rings, the target of dark and light bands that appeared when a curved and a flat piece of glass were pressed together. Grasping at straws, Newton theorized that the colors were caused by light particles undergoing "fits of easy reflexion and transmission."

No better theory was established until 1801, when Thomas Young (in his famous two-slit experiment) showed how two

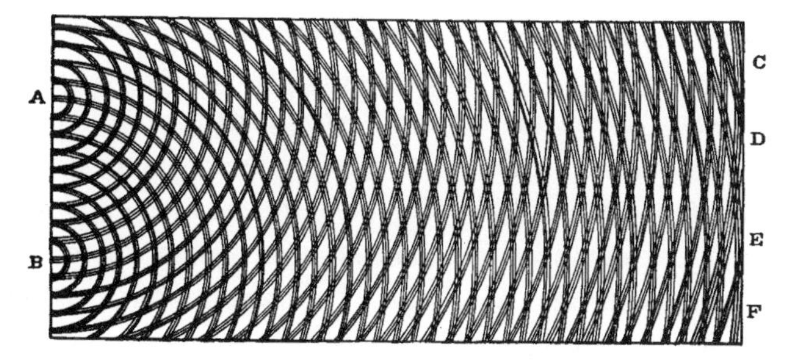

Thomas Young's interference pattern

overlapping light beams can interfere with each other, producing a similar pattern. The only way to explain this, Young proposed, was with waves. The lighter sections were produced when two wave crests overlapped, the darker sections when the crests were out of phase. After other confirming experiments, the wave theory came to be considered almost gospel, but it left a nagging question: What was doing the waving?

The answer that emerged was another of those imponderables: the "luminiferous aether," an ineffable something that pervaded everything—even the spaces between atoms. As rarefied as nothingness itself, aether was said to have the ability to vibrate and transmit light. More fundamentally, it promised an antidote for the celestial sailor's nightmare. Drifting through space, we cannot fix our position or velocity against the neighboring stars, for the stars are moving too. But everything could be measured against the aether.

In 1880, two years after his celebrated experiment at Annapolis, Michelson took a year's leave from the navy to study in Europe. Traveling with his family to Paris, where Margaret had gone to finishing school, he conferred with French physicists about a plan to measure the motion of the Earth against the aether. If he was right, a light beam sent in the same direction that the Earth was moving around the sun should be slowed a little by an aether wind. Proving so would be a matter of measuring light speed upwind and downwind and comparing the two. But that posed a problem. Each beam would have to bounce off a mirror, as in the Annapolis experiment, in order for the deflection to be observed. Any change in velocity from traveling one way would be canceled out in the other direction. (Swimming upstream and then down takes the same amount of time as swimming downstream and then up.)

But what, he proposed, if the beacons were sent out at right angles, one in the direction of the Earth's orbit and the other crossways? Now, as Michelson put it, one swimmer is "struggling upstream and back, while the other, covering the same distance, just crosses the river and returns. The second swimmer will always win, *if there is any current in the river.*"

Or in the case of the light beams, if there is an aether wind.

Moving on to Berlin later that year, he began assembling his apparatus. The handmade optics were expensive, but with the help of a colleague back home, Michelson got a grant from Alexander Graham Bell.

In the experiment, light from a lantern would be focused onto a half-silvered mirror, which would split the beam into two luminous "pencils," running in perpendicular directions. Traveling along two finely machined brass arms, each a meter long, they would ricochet off mirrors and come back together again. If the beams had moved at different speeds they would be slightly out of phase, with the crests of their waves not quite lining up.

The result would be an interference effect like the one Thomas Young had described: a pattern of dark and bright lines, or "fringes." Revolve the instrument ninety degrees, changing its orientation to the aethereal river, and the fringes should move. Taking into account the speed of the Earth against the aether and the wavelength of the light, he predicted a displacement of at least one-tenth of a fringe, something he was confident he could measure.

In so delicate an experiment, the slightest vibration might throw off the path lengths and spoil the results. ("So extraor-

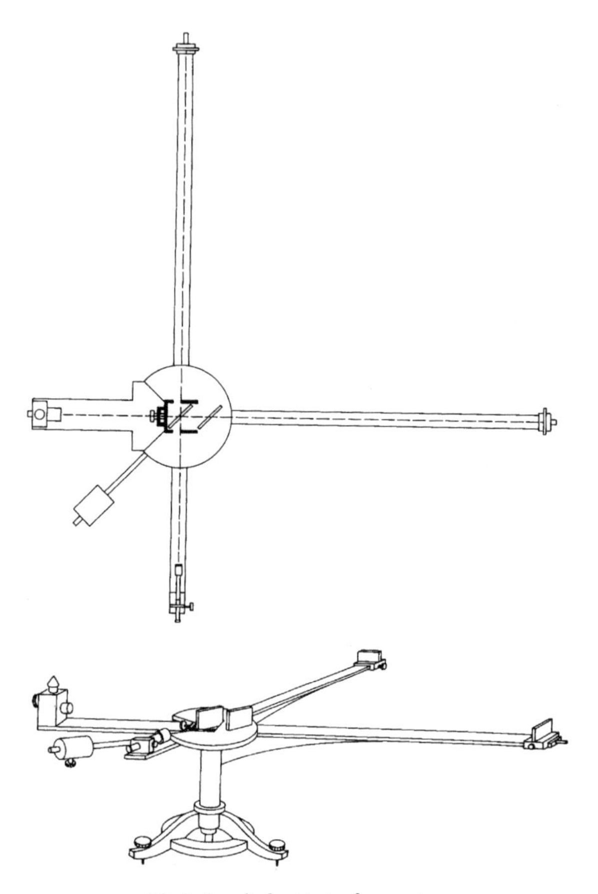

Michelson's first interferometer,
viewed from the top and from the side

dinarily sensitive was the instrument," he later noted, "that the stamping of the pavement about 100 meters from the observatory, made the fringes disappear entirely!") To keep the device—the interferometer—steady he anchored it to a stone pier. To minimize temperature differences, which might cause the brass arms to expand or contract, he covered them with paper boxes, and even tried surrounding the equipment with melting ice. The precautions were not enough. Even after midnight, the bustle of Berlin made it impossible to take a reading.

In search of quieter surroundings, he relocated to Potsdam and installed the equipment in the cellar of the Astrophysical Observatory. At first, as he rotated the device, he thought he saw a substantial fringe shift. Then he realized he was accidentally flexing the brass arms. He had the pivot remade to turn more freely and tried again.

Day after day he measured, turning his interferometer this way and that, but he could find no more than the tiniest shift—1⁄100 of a fringe—so slight that he could only dismiss it as experimental error. By now it was early April, when the Earth moved in the same direction as the whole solar system, increasing its speed against the aether, yet there still appeared to be no significant effect. Writing to his benefactor, Bell, in 1881, he reported the negative result. Michelson made clear that this shouldn't be taken as disproving the existence of the aether. There had to be aether. But maybe, as other physicists had suggested, the backdrop wasn't entirely fixed. Perhaps some of the aether in the vicinity of Earth was being dragged along in its journey around the sun. Traveling in the eye of a hurricane there would be no wind. Michelson's confidence was unshakable. "I have a very high respect for his abilities," Bell would later write, adding: "though I rather suspect from his manner that he has too."

MICHELSON'S only hope was that the aether drag was not complete, that enough of the celestial backdrop stayed put to provide a landmark to measure by. This possibility had been suggested earlier in the century by a French scientist, François Arago, who had tried to measure the velocity of starlight colliding with the Earth. Arago assumed, naturally enough, that the speed would vary depending on whether the orbiting planet was approaching or retreating from the light source. He mounted a prism on the end of a telescope, predicting that faster light beams would be bent more abruptly than slower ones. He was surprised to find that whatever the season the angles were the same.

Arago concluded that our eyes must be sensitive to only a small range of velocities, that the faster and slower rays were invisible. But his colleague Augustin-Jean Fresnel came up with a different explanation: while aether flows effortlessly through matter's molecular cracks, a tiny bit had become stuck in Arago's prism, carried along for the ride. That, he explained, would negate the effect Arago was seeking. When the Earth was approaching a star, its light would indeed strike the prism at a higher speed. But then it would be slowed a compensating amount by the aether trapped inside the glass. The effect would be true for any transparent medium, Fresnel proposed, and would depend on its index of refraction—a measurement of how much it slows and bends light. Aether drag would thus be noticeable in water but insignificant in air.

In 1882, after his sabbatical in Europe, Michelson left the navy and joined the faculty of the Case School of Applied Science in Cleveland, which had just opened its doors. During his first year, he measured the speed of light in a vacuum (almost dead-on at 186,320 miles per second). Then, with a man he had befriended on a train trip to Montreal, he began to rethink the aether experiment.

Edward Morley, a chemist at neighboring Western Reserve University, was as meticulous a scientist as Michelson. The two men agreed that it would be pointless to make another attempt to detect the Earth's absolute motion unless they could first confirm Fresnel's hypothesis—that the celestial backdrop is fixed in space with only pinches of aether dragged along by transparent objects. So slight an effect could be adjusted for. Improving on an experiment Fizeau had done, they pumped water through a loop of tubing and split a light beam so that one pencil moved with the current, the other against. They ultimately confirmed that there was indeed a small push and pull by the water. (Anachronistic aside: though they took this as confirmation of the aether drag hypothesis, the phenomenon is now explained as an effect of special relativity.)

It was in the midst of this experiment that Michelson fell apart. The reasons are obscure. He and his wife had been stuck in a bad marriage. He thought she talked too much at social gatherings, always trying to steal the show. She was bored with Cleveland, tired of her husband's late nights at the lab, or wherever he was. She complained that he took money from the household account to buy scientific equipment. When Michelson left for New York to be treated, Morley doubted that he would ever return to science.

Maybe that was wishful thinking. (Michelson had been treating Morley as shabbily as anyone.) Less than two months later he was back in the laboratory, ready to resume the experiment. There was another setback—in 1886 a fire destroyed the Case School, and Michelson had to move what was salvaged to Western. Finally, the next spring, the two men were ready for what they hoped would be the definitive

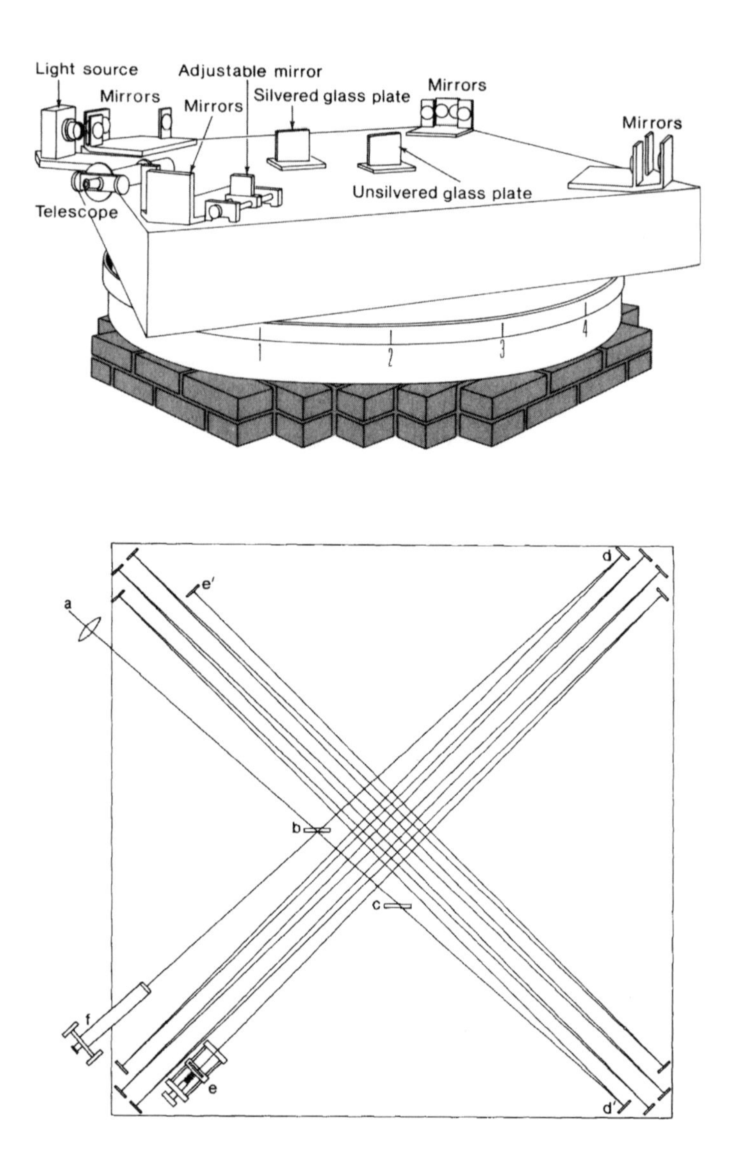

The Michelson-Morley experiment. The lower diagram shows the paths of the two light beams, which were extended by bouncing them back and forth between sixteen mirrors.

test, to determine, as Morley put it, "if light travels with the same velocity in all directions." Like Michelson, he assumed the answer was no.

This time even more care was taken to cushion the interferometer against the slightest vibration. The pieces were mounted on a sandstone slab, about five feet square and fourteen inches thick, which was attached to a wooden buoy, shaped like a doughnut and floating in a cast-iron trough of mercury. The trough itself was set on a concrete bed atop a brick platform. Four metal mirrors were set at each corner to reflect the light from an Argand lamp back and forth, increasing the path lengths—the one going with the Earth and the one moving across—to thirty-six feet. A wooden cover protected the optical instruments from the air. After carefully measuring and adjusting the distances between the mirrors—a calibration so precise that it required a screw with 100 threads to the inch—they began the experiment.

With a push of the hand, the interferometer was set slowly moving, one turn every six minutes, while Michelson walked alongside. Taking care to avoid touching the observing scope, he peered through the eyepiece at the interference fringes, calling out a reading to Morley at sixteen stations around the dial. Between July 8 and 12 they took observations both at noon and in the evening, and found no significant difference. The two swimmers returned at the same time.

They had intended to take samples during different seasons, to see if Earth's orbital motion made a difference, but there seemed little point. Fresnel must be wrong: a substantial amount of aether was indeed being dragged along with the planet, obscuring the effect. Measuring the absolute motion of the Earth would require carrying out measurements high above ground, maybe even in outer space.

Morley and another colleague, Dayton Clarence Miller, continued to look for aether using interferometers with even longer light paths. Miller even claimed to have detected the airy stuff with an experiment atop Mount Wilson, but he was apparently fooled by temperature fluctuations. In 1930 Michelson's own experiments on the mountain reconfirmed his original results.

It was not what he had wanted. By then he had remarried, sired a second family, and been honored with a Nobel Prize. But he had sought a deeper anchor: aether, "one of the grandest generalizations of modern science—of which we are tempted to say that it ought to be true even if it is not."

He died a year later, in 1931, just months after meeting Einstein, whose special theory of relativity had explained the true significance of Michelson and Morley's beautiful experiment: they had proved, contrary to their expectations, that there is no fixed backdrop of space, or even of time. As we move through the universe, our measuring sticks shrink and stretch, our clocks run slower and faster—all to preserve the one true standard. Not aether but the speed of light.

CHAPTER 9

Ivan Pavlov

Measuring the Immeasurable

Ivan Pavlov

We must painfully acknowledge that, precisely because of its great intellectual development, the best of man's domesticated animals—the dog—most often becomes the victim of physiological experiments. During chronic experiments, when the animal, having recovered from its operation, is under lengthy observation, the dog is irreplaceable. Moreover, it is extremely touching. It is almost a participant in the experiments conducted upon it, greatly facilitating the success of the research by its understanding and compliance.

—Ivan Pavlov

To hear him talk, you would have thought they were volunteers, these animals recruited for the research that would make Ivan Petrovich Pavlov a famous man. Lada, Lyska, and Zhuchka had common canine names. There were Pestryi (Spot), Laska (Weasel), Sokol (Falcon), Tsygan (Gypsy), Ryzhaia (Redhead), Pudel (Poodle), and Voron (Crow). There were Arleekin the Clown, Krasavietz the Beauty, Lyadi the Lady, Postrel the Fast One, Zloday the Thief, and Rogdi the Old Russian Prince. There were dogs named Baikal (after a Siberian lake) and Genghis Khan. And at the very beginning there was the one said to be Pavlov's favorite, a setter-collie mix he called Druzhok, for Buddy or Little Friend.

They had it better than animals in other physiology labs that still employed the "acute" experiment: cutting open and sacrificing a living animal to observe the anatomical ticking. To Pavlov this was like smashing a watch with a mallet to see how it ran. Beginning with his pioneering studies of the mammalian digestive system, still at the core of gastroenterology, he favored the "chronic" approach: while the dog was under anesthesia, its stomach, esophagus, or salivary glands would be altered so fluids could be collected and analyzed. Pavlov became known as one of the most skilled surgeons in Europe, and he operated under antiseptic conditions better than those in many hospitals. Only when the animal had fully recovered would the observations, extending over months or years, begin.

By the early 1900s, when his interest had turned to the nervous system, the symbiosis was complete. In return for room and board, the dogs became experimental subjects, and also mascots. Between sessions in the laboratory, they were taken for walks on the institute grounds. Sometimes to clarify a point of physiology, Pavlov resorted to acute experiments, but only with regret. "When I dissect and destroy a living animal, I hear within myself a bitter reproach that with rough and blundering hand I am crushing an incomparable artistic mechanism. But I endure this in the interest of truth, for the benefit of humanity." In a world where animals were hunted for recreation and slaughtered for food and leather, he felt justified in using a few for the pursuit of knowledge.

It was the usual answer one gave to the antivivisectionists, who were a part of the scene in Russia, as they are throughout the world today. From their perspective Pavlov's experiments were anything but beautiful. Even a dog owner unperturbed by foie gras on a restaurant menu or the fate of a laboratory mouse might wince at the surgical descriptions. The consolation is the knowledge that was gained. With its crisp logic and elegant design, the work with Pavlov's dogs opened a passage to a world that had seemed as remote as the farthest star: the inside of the brain.

HE HAD intended to become a priest, like his father, in the Russian Orthodox church. Then he discovered Darwin. It was the late 1860s, and Ivan and his brother, Dmitry, were studying at the seminary in Ryazan, where the Pavlovs lived. Early in the morning, the story goes, Ivan would sneak into the village library to read the recent Russian translation of *On the Origin of Species* as well as George Henry Lewes's *Physiology of Common Life,* with its maplike diagrams of internal organs, and Ivan Sechenov's *Reflexes of the Brain,* a radical exercise in pure materialism arguing that the mind was nothing more than an exceedingly complex machine.

Sechenov proposed that every human behavior, from a sneeze to a decision to read a book, consists of reflexes—muscular movements triggered by signals registered by the senses. "Absolutely all the properties of the external manifestations of brain activity described as animation, passion, mockery, sorrow, joy, etc., are merely results of a greater or lesser contraction of definite groups of muscles," he wrote, "which, as everyone knows, is a purely mechanical act." Even when a thought pops into the head unbidden, it is the product of a reflex, he insisted, the evoking of a buried memory by subtle environmental cues. "The time will come," Sechenov declared, "when men will be able to analyze the external manifestations of the functioning of the brain as easily as the physicist analyzes now a musical chord or the phenomena of a freely falling body."

These were invigorating ideas for the son of a priest. Under the reign of Czar Alexander II, a penumbra of enlightenment was crossing the Russian steppes. Books and journals that would have been banned under his father, Nicholas I, were arriving at the library, where a crowd gathered at the doors waiting for them to open, pushing and shoving to get in. To beat the rush Pavlov would sometimes arrange for a worker to leave a window open.

Fascinated by the notion that the animal organism could be understood scientifically, he left the seminary in 1870 to study in Saint Petersburg. Dmitry soon joined him, and they both studied chemistry under Mendeleyev, who was devising his periodic table of the elements. Ivan, however, concentrated on physiology, eventually earning a doctorate of medicine for experiments on how the canine nervous system controlled blood pressure and the pumping of the heart. In 1891 he was appointed head of physiology at the newly formed Institute for Experimental Medicine, where he used his surgical techniques to map out the cascade of functions—a "complex chemical factory," he called it—through which food was processed and absorbed by the body.

Even before a morsel was placed on a dog's tongue, the flow of saliva began: water for dilution mixed with mucin to lubricate the food for its passage to the stomach, where a batch of "appetite juice" was already being prepared. There and later in the duodenum, specialized nervous sensors analyzed the food, signaling the body to secrete the proper recipe of gastric fluids needed to digest bread, milk, meat, or whatever the dog was having for dinner.

Salivation, Pavlov noticed, also served another function. Give the animal a taste of something unsavory—mustard oil, mild acid, or salt—and saliva still flowed. But it consisted mostly of water to protect the tongue and wash out the noxious substance. In this case there were no gastric secretions. The organism somehow "knew" they were unnecessary.

To measure the amount and the composition of the saliva, Pavlov subjected the dogs to a minor operation. While the animal was anesthetized, the opening of the duct leading from one of the salivary glands was moved to the outside of the chin or cheek and secured with a few stitches. Once the incision had healed, fluid was collected and analyzed. Pebbles of quartz, he found, produced hardly a drop, while sand released water so the dog could wash it out. By the same physio-logic, a dog actually drooled more at a piece of dry bread than a savory chunk of meat. Each reflex was fine-tuned by evolution to synchronize the animal with its environment.

Scenes from the Institute of Experimental Medicine

"Every material system can exist as an entity only so long as its internal forces, attraction, cohesion, etc., balance the external forces acting upon it," he later wrote. "This is true for an ordinary stone just as much as for the most complex chemical substances, and its truth should be recognized also for the animal organism. . . . Reflexes are the elemental units in the mechanism of perpetual equilibration."

In 1904 Pavlov won a Nobel Prize for his work on the physiology of digestion, an honor he was almost denied when a rival laboratory discovered that he had missed an important component of the system: hormones. "It is clear that we did not take out an exclusive patent for the discovery of truth," he said fatalistically. It was around this time that he decided to leave digestion to others and concentrate on what he called the highest parts of the nervous system.

PAVLOV had noticed that for salivation to occur, it was not actually necessary for food to enter an animal's mouth. The smell, the appearance of the bowl, even the creaking of a door hinge at dinnertime might be enough to set off the reaction. "Psychic secretions," he called them.

Unlike the inborn reflexes—the instincts—these acquired or "conditional" responses could be modified. Show a dog a piece of meat and then take it away. Do this several times and the animal will salivate less and less. There has been an "inhibition" of the reflex. A taste of meat, bread, or even, paradoxically, noxious acid restores ("disinhibits") the reaction. Just as evolution acting over eons molds a species to its environment, experience acting over a lifetime molds an individual organism to the details of its particular locale. It has evolved the ability to learn.

Early on, Pavlov was tempted to interpret these phenomena psychologically, imagining what thoughts might be unfolding on the dog's inner screen. The animal stopped drooling after repeated showings of meat because it had become jaded, cynical, as though "convinced of the uselessness of its efforts." But why then would the revolting touch of acid bring salivation back? What could the dog be thinking?

This, Pavlov came to believe, was the wrong question. "Indeed, what means have we to enter into the inner world of the animal!" he later declared. "What facts give us the basis for speaking of what and how an animal feels?" The same, he poignantly observed, applies to people. "Does not the eternal sorrow of life consist in the fact that human beings cannot understand one another, that one person cannot enter into the internal state of another?"

The line between the mental and the physical was beginning to blur. When a scientist studied how blood pressure rises and falls or pancreatic juices flow, Pavlov noted, he spoke in purely material terms. "But now the physiologist turns to the highest parts of the central nervous system, and suddenly the character of his research sharply changes. . . . He begins to make suppositions about the internal state of animals, based on his own subjective state. Up to this moment he had used general scientific conceptions. Now he changes front, and addresses himself to foreign conceptions in nowise related to his earlier ones, to psychological ideas. In short, he leaps from the measurable world to the immeasurable."

It was time to concentrate on the objective. Whether the salivary glands were fired by receptors in the tongue or in the eye, nose, or ear, the result was the same: signals from the environment were eliciting a physiological reaction.

THE IDEA that organisms, their brains included, are biological machines was broached by Descartes in the seventeenth century, but he allowed that there was something special about his fellow humans. Although our bodies are purely mechanical, constrained to obey the principles of physics, our brains are inhabited by a higher presence, the mind. By Pavlov's time, Darwin's discoveries had made this kind of dualism tricky to maintain. The brain presumably evolved along with the rest of the body, but how could the material tugs of natural selection act on the ghostly mind? William James described the problem in 1890 in *The Principles of Psychology:* "The self-same atoms which, chaotically dispersed, made the nebula, now, jammed and temporarily caught in peculiar positions, form our brains; and the 'evolution' of the brains, if understood, would be simply the account of how the atoms came to be so caught and jammed."

Some philosophers went so far as to propose that each atom of matter is shadowed by an atom of consciousness—"primordial mind-dust" that was carried along as the cosmos unfolded and species evolved. James explained their reasoning: "Just as the material atoms have formed bodies and brains by massing themselves together, so the mental atoms, by an analogous process of aggregation, have fused into those larger consciousnesses."

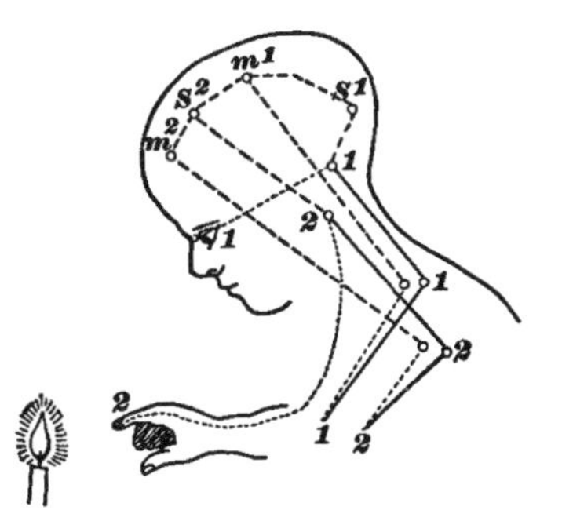

A baby acquiring an avoidance reflex to fire. Diagram from William James, *Principles of Psychology*

Running on a parallel track, every chemical action in the brain was said to be mirrored by a mental action, with neither exerting control over the other. Thomas Henry Huxley had put it like this: "The soul stands related to the body as the bell of a clock to the works, and consciousness answers to the sound which the bell gives out when it is struck." When we "decide" to move a finger, that is an indication, not the instigator of the event. "The feeling we call volition," Huxley proposed, "is not the cause of a voluntary act, but the symbol of that state of the brain which is the immediate cause of that act." (A century later, the physiologist Benjamin Libet claimed to have demonstrated just that.)

We are, in other words, conscious automata. James disapprovingly described the implications:

> If we knew thoroughly the nervous system of Shakespeare, and as thoroughly all his environing conditions, we should be able to show why at a certain period of his life his hand came to trace on certain sheets of paper those crabbed little black marks which we for shortness' sake call the manuscript of *Hamlet.* We should understand the rationale of every erasure and alteration therein, and we should understand all this without in the slightest degree acknowledging the existence of the thoughts in Shakespeare's mind. The words and sentences would be taken, not as signs of anything beyond themselves, but as little outward facts, pure and simple. In like manner we might exhaustively write the biography of those two hundred pounds, more or less, of warmish albuminoid matter called Martin Luther, without ever implying that it felt.

Pavlov didn't linger long on such metaphysical matters. Whatever might be happening inside a dog's mind could be approached only from the outside, objectively. "The naturalist must consider only one thing: what is the relation of this or that external reaction of the animal to the phenomena of the external world?"

These signs, he was quick to learn, need bear no inherent relation to what they signify. It is natural that a dog's mouth would water at the smell of meat, though that too seemed to be a learned response. (A puppy still imbibing its mother's milk may turn up its nose at a hamburger.) But by presenting meat at the same time as another stimulus, the experimenter could train the animal to salivate at the flash of a light, the rotation of an object, the touch of a hot or cold probe to the skin, the ticking of a metronome, or the sound of a buzzer, whistle, tuning fork, or horn. (Pavlov hardly ever used a bell.) There is no reason evolution would anticipate such arbitrary pairings. But under the circumstances they became meaningful to the dog's survival.

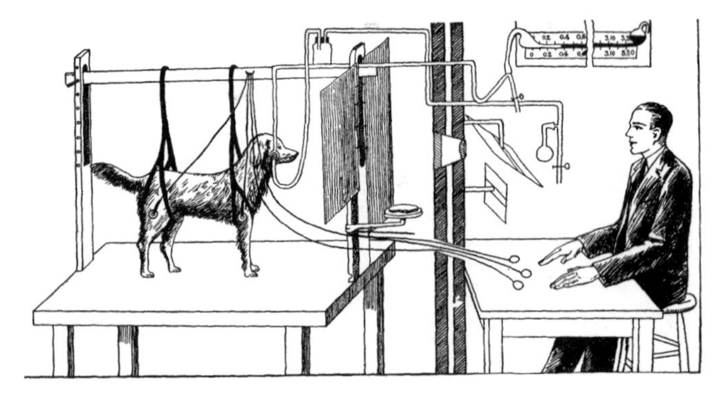

Training a dog to salivate when two mechanical stimulators prick its skin

The same was true of the defensive salivary reaction. Once the dog had sampled dilute acid dyed black with India ink, it would drool protectively at the sight of black water. But after it had sampled the harmless solution several times, the reflex disappeared, only to be restored with another taste of the acid.

So malleable were the neural connections that they could be plugged and unplugged like cables in a telephone switchboard. With enough training, a positive stimulus like a piece of meat could be linked with an obnoxious one. Instead of recoiling at an electric shock the dog would drool.

As his technique became more practiced, Pavlov's laboratory began investigating the canine sense of time. After a dog was trained to salivate at a flash of light, the delivery of the stimulus was postponed by three minutes. Before long, the dog learned to anticipate the delay. Three minutes after the signal, the animal's mouth would water.

In other experiments time itself became the stimulus. Give a dog food every thirty minutes. When the feedings are suspended, it will continue to salivate robotically on the half hour. "I am convinced," Pavlov declared, a bit grandiosely, "that directly along this path of exact experimentation lies the solution of the problem of time, which has occupied philosophers for countless generations."

So precise was their neural machinery that the dogs could even be conditioned to discriminate between an object rotating clockwise and one rotating counterclockwise, between a circle and an ellipse, between a metronome beating 100 times per minute versus 96 or 104. They could distinguish between adjacent notes on a musical scale, between C and F played in any of five different octaves on an organ, and among different shades of gray.

For experiments like this, the context was essential. If a dog learned a new reflex while it was sitting on the floor, the experiment might fail if it was repeated on a table, or by a different experimenter. Distractions had to be carefully avoided. "Footfalls of a passer-by, chance conversations in neighboring rooms, the slamming of a door or vibration from a passing van, street-cries, even shadows cast through the windows into the room, any of these casual uncontrolled stimuli falling upon the receptors of the dog set up a disturbance in the cerebral hemispheres and vitiate the experiments."

Pavlov's dogs were as temperamental as Michelson's interferometer. Determined to control every possible variable, he commissioned the construction of a "Tower of Silence" modeled on seismological laboratories. The building was surrounded by a straw-filled moat to dampen vibrations, and its first and third floors each had four soundproof observation chambers isolated by corridors and the unoccupied floor in between. Experimenters observed the dogs remotely through periscopes, giving the impression, one visitor reported, of "a submarine ready for battle."

"Pavlov's physiology factory," as the historian Daniel Todes called it, was a sign of what experimental science would become. Under Pavlov's direction teams of researchers tested hypotheses on hundreds of different dogs. What emerged was not, perhaps, a single beautiful experiment but a suite of them. Still, one was so surprising that it stands above the rest.

Pavlov and his collaborators had already shown that a dog had basic musical abilities. Trained to salivate to a specific chord, say A-minor, it would also react—albeit more weakly—to each individual note. Pushing still further, the researchers began testing the animal's ability to recognize simple melodies.

When four notes were played in ascension, the dog was given a bit of food.

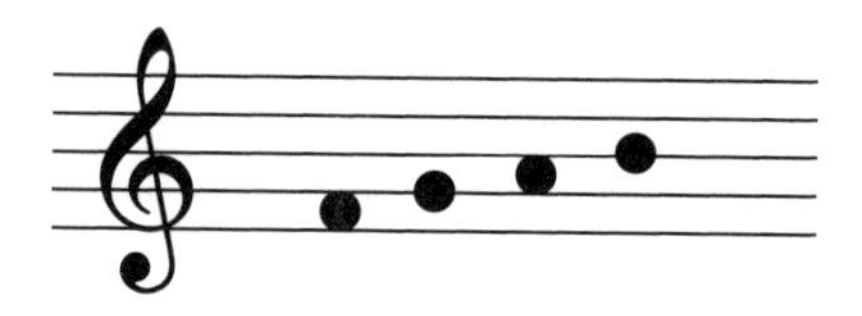

When the same notes were played in descending order, there was no reinforcement.

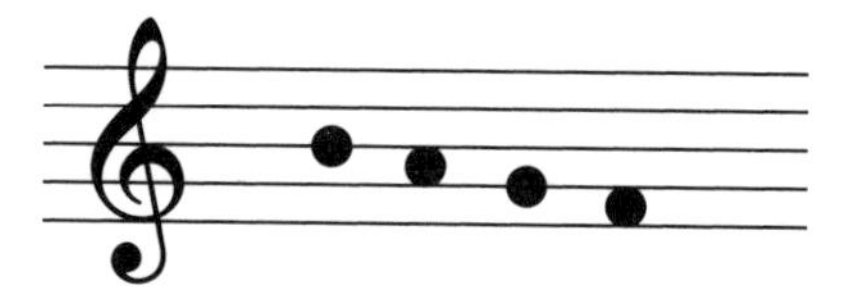

The animal quickly learned to tell one sequence from the other. But how, Pavlov wondered, would it respond when it heard the twenty-two other possible combinations of the same notes?

The melodies were played and the spittle collected. The dog had categorized the scales into two equal groups depending on whether the pitches were predominantly rising or falling. It's not too much of a stretch to say that the animal had formed a rudimentary concept. This kind of pattern recognition, Pavlov came to believe, was the root of what he himself did as an experimental scientist.

"The movement of plants toward the light and the seeking of truth through a mathematical analysis—are these not phenomena belonging to the same order? Are they not the last links in an almost endless chain of adaptabilities which appear everywhere in living creatures?"

Like many scientists with a powerful theory, he got carried away at times, trying to explain his dogs' personalities and even human neurosis as bundles of conditional reflexes. In the United States, John B. Watson and B. F. Skinner developed the psychology of behaviorism, in which everything mental was reduced to stimuli and responses. The result was two clashing visions of the future: Skinner's novel *Walden Two* describes a utopia brought on through behavioral engineering, while in Aldous Huxley's *Brave New World*, the same tools are used by the state to impose a crushing dictatorship. Neither has come to pass. More recently the metaphor of the computer has given scientists a more nuanced way to think about thinking, but Pavlov's fundamental realization has endured: the brain and nervous system form a precise, highly adaptable living machine.

Later in his life Pavlov's students gave him an album with photographs of forty of his dogs. A copy was tracked down

Pavlov's dogs

Monument to a Dog

in Saint Petersburg by a scientist at Cold Spring Harbor Laboratory who was using Pavlovian conditioning in fruit flies to identify genes involved in long-term memory. He named the various mutants—"Pavlov's flies"—after the famous animals.

In 1935, *Monument to a Dog,* an ornate fountain, was built on the grounds of the institute. At the core is a pedestal with a large canine sitting on it with bas reliefs of laboratory scenes and quotations from Pavlov: "Let the dog, man's helper and friend since prehistoric times, offer itself as a sacrifice to science. But our moral dignity obligates us to ensure that this always occurs without unnecessary pain."

Around the top are busts of eight canines, water pouring from their mouths as they salute in salivation.

CHAPTER 10

Robert Millikan

In the Borderland

Robert Millikan

> We have actually touched the borderland where matter and force seem to merge into one another, the shadowy realm between Known and Unknown which for me has always had peculiar temptations. I venture to think that the greatest scientific problems of the future will find their solution in this Border Land, and even beyond: here, it seems to me, lie Ultimate Realities, subtle, far-reaching, wonderful.
>
> —William Crookes, 1879

ON A SATURDAY morning in January, in search of the last piece of equipment I needed to persuade myself that electrons exist, I set out for the "Black Hole," a postapocalyptic junkyard ("Everything goes in and nothing comes out") in Los Alamos, New Mexico. Run by Edward B. Grothus, a former bomb maker and now aging peace activist, the warehouse—converted from an old grocery store—is packed floor to ceiling with oscilloscopes, signal generators, Geiger counters, vacuum pumps, centrifuges, ammeters, ohmmeters, voltmeters, cryogenic storage vessels, industrial furnaces, thermocouples, barometric gauges, transformers, typewriters, ancient mechanical calculators—more than seventeen thousand square feet of electronic and mechanical detritus cast off over the years by the national laboratory where the Manhattan Project began.

Over the years I had acquired on eBay most of what I'd need to repeat the classic experiments: J. J. Thomson's 1897 demonstration that electricity is a form of negatively charged matter, followed thirteen years later by Robert Millikan's triumphal oil-drop experiment, isolating and measuring the charge of individual electrons. Combing the dark aisles of the Black Hole, I finally spotted what I'd been looking for: a Fluke 415B High Voltage Power Supply. Reaching over my head, I carefully freed the long gray chassis from the middle of a stack—it weighed thirty pounds—and lowered it to the concrete floor. Built in the 1960s and operated by vacuum tubes, it appeared to be in perfect condition. Dragging it to the back of the store, where miles of coaxial cables hung snakelike from hooks or lay coiled on the floor, I found one that fit the output connector and made my way to the cash register.

Ed never seems to actually want to sell anything. He'd rather tell you about his plan to erect a pair of granite obelisks to surprise alien archeologists after the coming holocaust, or about his First Church of High Technology, where he performs a "critical mass" on Sundays. By the time some customers tracked him down in the depths of his lair, he was in a cantankerous mood. "Two hundred fifty dollars for that," he said—about ten times what I'd been expecting. I tried to reason with him. There was one exactly like it on eBay for $99. But Ed is not a man to bargain. Disappointed, I dragged the unit back to its resting place, where it is probably still sitting, and left with just the cable. Stopping at the public library, next to Fuller Lodge, where Oppenheimer and the other nuclear physicists partied and dined, I signed on to the Internet and bought the other power supply. Two weeks later it arrived and I was ready to begin.

IN 1896, Robert Andrews Millikan, a young physicist fresh out of Columbia University with a PhD, found himself at a lecture in Berlin where Wilhelm Roentgen was showing pictures he had taken of the bones inside a hand. The occasion was a January meeting of the German Physical Society, and Millikan felt such childlike wonder that he later misremembered the talk as occurring on Christmas Eve.

Just two years earlier, in the United States, he had heard the great Albert Michelson speculate that physics was all but over. The laws of motion and optics were set firmly in place, and Maxwell's equations had drawn tight the threads Faraday and his generation had spun between electricity and magnetism. Heinrich Hertz had gone on to verify Maxwell's theory, showing that radio waves can be reflected, refracted, focused, and polarized—that they are just a kind of light. But here was a new, entirely unexpected phenomenon. X-rays.

The prevailing wisdom, Millikan was happy to realize, had

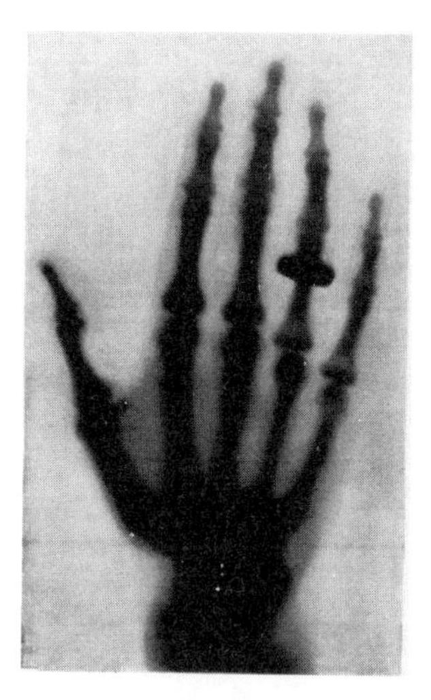

Roentgen rays
look inside a hand

been wrong. "We had not come quite as near sounding the depths of the universe, even in the matter of fundamental physical principles, as we thought we had."

Roentgen had made his astonishing discovery while investigating the glowing spot that appears at the end of an evacuated glass "discharge tube" when a large enough voltage is applied across two metal plates inside—a negatively charged cathode and a positively charged anode (names that had come from Faraday). Traveling through the rarefied air, these cathode rays were puzzling enough. If a tube was designed with an obstruction inside—William Crookes, a chemist and spiritualist, used a Maltese cross—its shadow would appear on the fluorescing glass, a clue that the rays moved bulletlike in straight lines. If he held a magnet near the tube, the beam would sway to one side. Mount a gemstone inside and it would fluoresce. The rays also seemed to have substance,

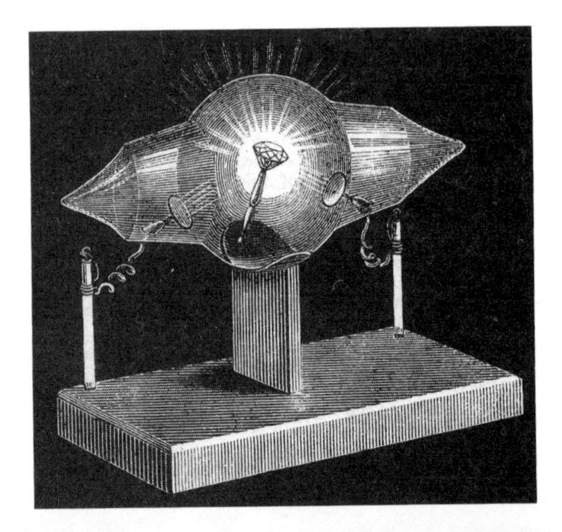

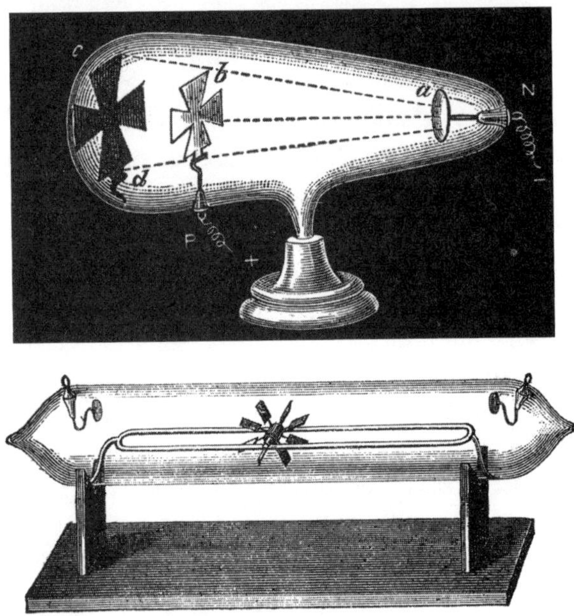

Crookes tubes: cathode rays light up a diamond, project a shadow of a Maltese cross, and move a paddle wheel along a track.

turning the vanes of a tiny paddle wheel. "A fourth state of matter," Crookes claimed—solid, liquid, gaseous, and radiant.

What Roentgen found was even weirder: if the beam struck the end of the tube with enough force, it unleashed a different kind of radiation—powerful enough to penetrate flesh. Less than a year later Henri Becquerel in Paris discovered another form of penetrating rays emanating from lumps of uranium, passing through an opaque shield and leaving their mark on a photographic plate. Both kinds of radiation, it soon was learned, could ionize a gas, giving it an electrical charge. We know now that they do this by knocking electrons off atoms.

Returning from Europe to take a job at the University of Chicago, where Michelson now reigned, Millikan watched from afar as some of Europe's greatest scientists explored the new physics. At the Cavendish Laboratory in Cambridge, England, J. J. Thomson showed that the beams could be repelled not just by magnets but by strong electrical fields.

Hertz himself had tried and failed at the experiment, in which a beam travels between parallel plates inside an evacuated tube. When the plates were charged with a battery, the beam didn't budge. Hertz took this to mean that the rays were an immaterial disturbance of the aether. (The lesson of Michelson-Morley was still sinking in.)

Thomson suspected that Hertz hadn't pumped enough air from the tube—that the lingering molecules were shorting out the plates as surely as if they had been rained on. With a better vacuum, he was able to nudge the beam toward the positive pole—a strong indication that cathode rays were made of negatively charged matter. Particles of electricity. Electrons.

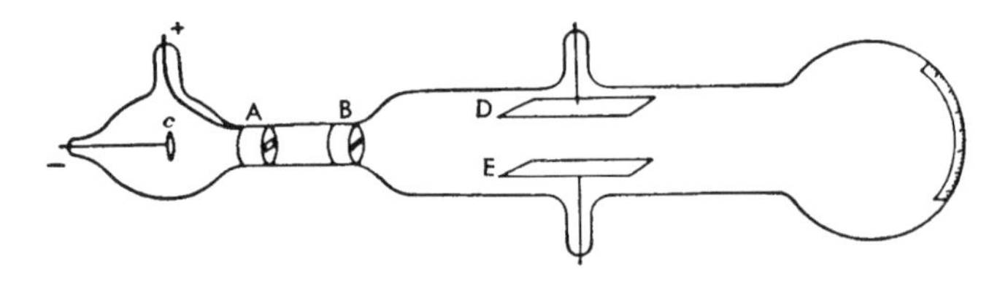

J. J. Thomson experiment. Cathode rays are emitted at C, pulled through the positively charged anode (A) then pass through slit B and between plates D and E before leaving a spot on the end of the tube. Charging the plates causes the beam to move.

I HADN'T meant to buy my own Thomson apparatus but its beauty was impossible to resist: the simple wooden frame cradling the bulbous, pointed vacuum tube, the large copper Helmholtz coils (named for the German physicist Hermann von Helmholtz) standing at either side. With the spacing between them equal to their radius—fifteen centimeters—they bathe the tube in a uniform magnetic field. The device was made in Germany for use in physics classes, and the grayish crackled finish on the electrical junction box dated it probably to the 1960s.

The manual was not included, just a heavy piece of drawing paper on which someone had sketched with colored pencils a wiring diagram: the filament required 6.3 volts to heat the metal cathode and boil off electrons, which would be accelerated by a much larger positive voltage on the anode. A third source of current would energize the Helmholtz coils. I hooked up the wires to my power supply and turned out the lights.

It was an eerie sight. As I slowly increased the anode voltage, a greenish apple-shaped haze gathered around the cathode, growing larger and fatter until suddenly, a hair above 160 volts, a blue ray of light shot straight up from the stem

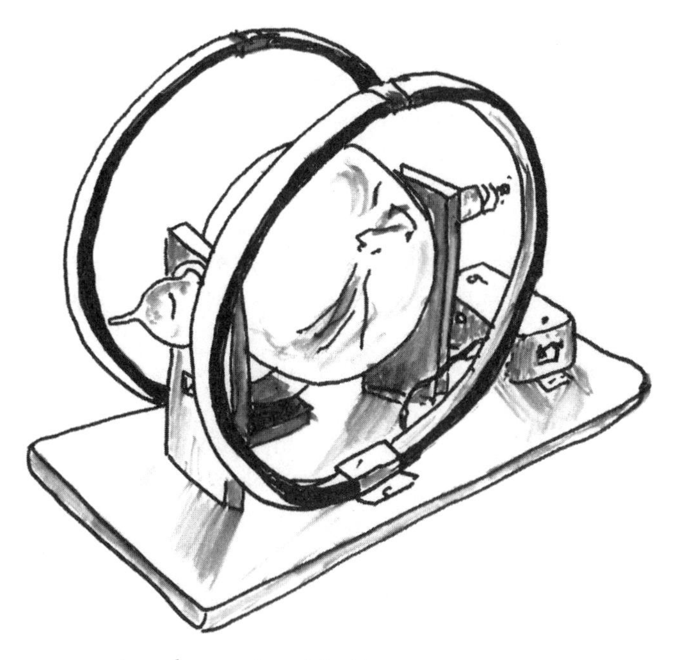

A modern version of the Thomson apparatus.
Drawing by Alison Kent

and struck the top of the glass. The genie in the bottle. How spooky this must have been for Crookes and the other cathode-ray pioneers. Some thought they were seeing ectoplasm. Ghost stuff. Holding a bar magnet to the glass, I made the genie writhe. The black pole beckoned the beam toward me, the red pole pushed it away.

The next step was to energize the coils. As I turned up the knob, the beam slowly bent until—at 3.5 volts, 0.76 amperes—it abruptly dived clockwise and formed a glowing circle inside the tube. While the anode was trying to pull the electrons straight upward, this magnetic wind was blowing them to the side—a perpendicular struggle whose outcome,

Thomson had realized, depends on both the mass of the particles and their charge. His experiment can't tell you either value alone (lightweight particles with a tiny charge would act the same as heavier particles with a larger charge), but it does give you their ratio.

I plugged my numbers—the voltage on the anode, the current in the coils, the radius of the glowing circle—into his equation and did the arithmetic: 2.5×10^8 coulombs of charge per gram. (A coulomb, named in honor of the French scientist Charles-Augustin de Coulomb, is approximately the quantity of electricity flowing each second through a 100-watt bulb.) My result was 50 percent larger than the accepted value, but at least I got the right number of zeros.

More important is what Thomson went on to show: that it didn't matter what kind of gas was in the tube or what metal he used for the cathode. The ratio remained unchanged. The rays were all made from the same stuff.

And what strange stuff it was. The ratio of charge to mass had already been measured for the hydrogen atom—the lightest of the elements—as it migrated between the poles of an electrolytic cell. The value for the electron was about a thousand times greater. Either it had an enormous charge, or as Thomson suspected, it was vastly smaller than an atom. His instincts told him he had discovered something almost unthinkable: a subatomic particle.

IT WAS 1906 and Millikan was feeling like a has-been—a decade at Chicago and still an assistant professor. He considered himself an effective teacher, and his textbooks were selling. But he was disappointed that at age thirty-eight, rather old for a physicist, he had made no important discoveries.

He knew that Thomson's experiment, impressive as it was, hadn't clinched the case. For all anyone knew, electrons came in a slew of charges and sizes all yielding the same ratio. Thomson had just assumed they were identical. In the face of this uncertainty, the Germans remained particularly skeptical, clinging to the belief that electricity was an aethereal wave. The only way to break the logjam would be to measure one of the numbers in Thomson's ratio—either the mass or the charge of the electron.

Millikan began by repeating an experiment in which a scientist in Thomson's lab at the Cavendish had timed how quickly a charged mist of water vapor—one that had been ionized with X-rays or radium—settled to the bottom of a closed container. Above and below the cloud were metal plates connected to the poles of a battery. By observing the effect of the electrical field on the speed of the cloud's descent, you could calculate its total charge. Divide that by your guesstimate of how many charged particles were in the cloud and you could rough out an average value for the electron.

The technique, which involved a device called a Wilson cloud chamber, was rife with uncertainty and assumptions. The vapor was continually evaporating, leaving the top edge of the cloud so irregular and indistinct that tracking its motion was an exercise in frustration. Millikan cranked up the voltage, hoping he could hold the target steady—suspended "like Mohammed's coffin" between positive and negative. Then he could measure the rate of evaporation and account for it in his calculations.

Instead he flicked on the switch and blew the cloud away. The experiment was a failure . . . or so it seemed until he noticed that a few individual water drops remained hanging

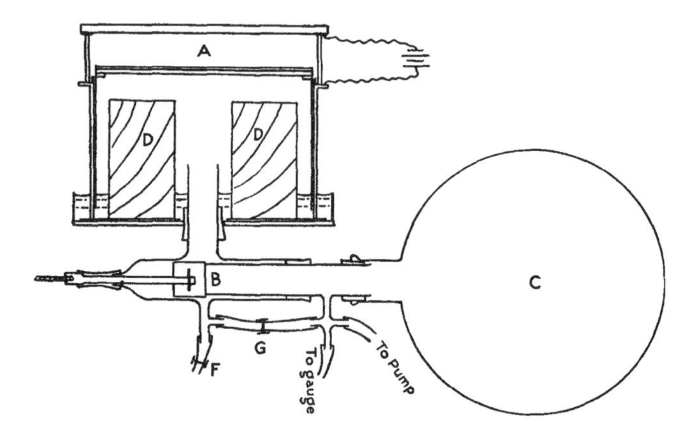

Wilson cloud chamber. Opening valve B causes a vacuum (C) to suck down the floor beneath chamber A, which is filled with moist air. The expansion of the volume causes a cloud to form.

in the air, just the right weight and charge so that the downward pull of gravity was offset by the levitating oomph of the electrical field.

This, he realized, would make for a more decisive experiment. Instead of studying the mass behavior of a whole cloud of drops, he could observe them one by one. Peering through a small telescope set up two feet away, he would pick a drop hovering in suspension and then suddenly turn off the voltage. Stopwatch in hand he timed the fall between the hairlines of his eyepiece. Hour after hour he recorded the data, comparing the estimated weight of a drop with how much charge was required to keep it afloat. The answer, Millikan reported, was always "1, 2, 3, 4, or some other exact multiple of the smallest charge on a droplet that I ever obtained." Charge indeed seemed to come in uniform portions—what he reckoned to be 1.55×10^{-19} coulombs.

In September 1909 he traveled to Winnipeg to present the results—he still considered them preliminary—to a meeting of the British Association for the Advancement of Science. Thomson himself gave the presidential address, and Ernest Rutherford, who had just won a Nobel Prize, lectured on the state of atomic physics, noting that for all the recent successes "it has not yet been possible to detect a single electron." Then Millikan, who wasn't even on the agenda, surprised everyone by reporting that he had come close to doing just that.

On the train back home he thought about how he might make a more persuasive case. Because of evaporation each water drop's lifetime was measured in seconds. How much better it would be if he could follow a single drop for minutes or even hours, adjusting the voltage and buffeting it up and down. As he was gazing out at the plains of Manitoba, the answer, he later said, came in a flash.

After arriving in Chicago, he asked Harvey Fletcher, a doctoral student who had been looking for a thesis problem, to see if the droplet experiment could be done with something less evanescent than drops of water. Purchasing a perfume atomizer and watch oil at a local drugstore, Fletcher began assembling the equipment: two round brass plates, the top one with a hole drilled at the center, mounted on a lab stand and illuminated from the side by a bright light. He sprayed a mist of oil above the apparatus and watched through a telescope. "I saw a most beautiful sight," he later recalled:

> The field was full of little starlets, having all the colors of the rainbow. The larger drops soon fell to the bottom, but the smaller ones seemed to hang in the air for nearly a minute. They executed the most fascinating dance.

By the next morning Fletcher had wheeled in a large bank of batteries capable of producing one thousand volts and connected them to the brass plates. Turning on the current, he watched with excitement as some of the droplets were pushed slowly upward while others were pulled down, the friction from the tiny nozzle of the atomizer having given them negative or positive charges. When Millikan saw how well the plan was working, he was elated. He and Fletcher refined the setup and spent nearly every afternoon for the next six months taking data.

DESIGNED and crafted by the Philip Harris Company of Birmingham, England, my setup was a streamlined version of Millikan's. But the idea was the same. The brass plates were mounted inside a three-legged Plexiglas platform that stood on a dark hardwood base measuring about fifteen by twenty inches. Off to one side was the lighting source: a

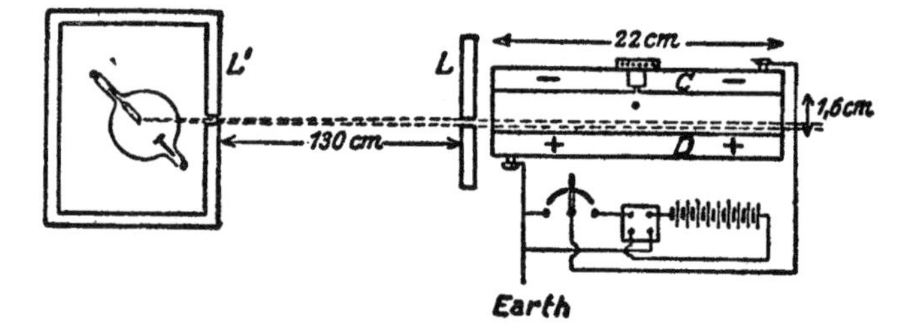

Early version of the Millikan oil-drop experiment. Droplets fall through the pinhole and into the space between brass plates C and D, which are connected through a switch to a battery. To the left is an X-ray source used to knock electrons off the drops and change their charge.

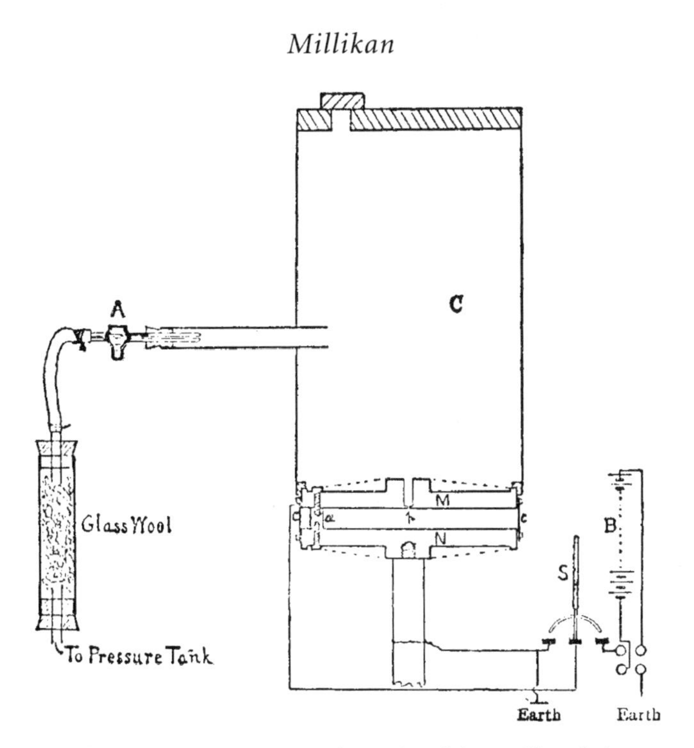

A later version. A commercial atomizer (A) uses filtered air to spray oil into chamber C from which an occasional drop makes its way through the pinhole in the top plate (M).

cylindrical metal housing, painted the familiar laboratory gray, with a lens to concentrate the glow. The British-sized bulb was missing, but I was able to substitute an ordinary halogen lamp powered by an old Lionel train transformer.

For peering between the plates at the dancing drops there was a telemicroscope (a cross between a telescope and a microscope) fitted with a crosshatched measuring reticule, and a knife switch for applying the electricity. *Up* sent power to the plates ("Do not exceed 2,000 volts," warned the black

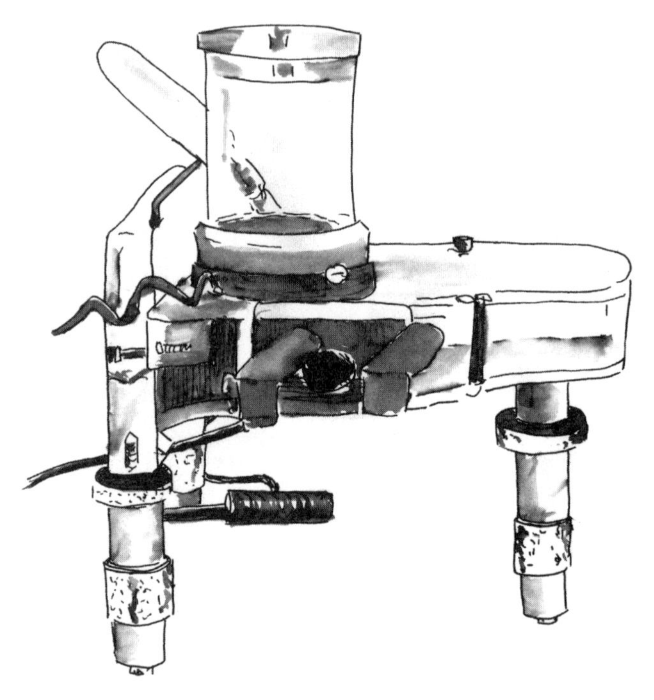

Philip Harris Co. Millikan apparatus.
Drawing by Alison Kent

Bakelite instruction panel.) *Down* shorted them together and dispelled the charge. After disassembling the parts to clean out dust and the accumulated oil of a thousand student experiments, I was ready for my first run.

I loaded a perfume atomizer with ordinary mineral oil and sprayed it into the chamber above the top brass plate. Then I waited for a few droplets to fall through the tiny hole. They looked more like dust motes in a shaft of sunlight than like little stars. But the effect was hypnotizing. I'd pick out one that was falling straight and slow and switch on the plate voltage. If it suddenly began moving upward I knew that it carried a charge. Flipping the knife switch up and down and adjusting the voltage, I'd time the drops as they rose and fell between the hairlines in the eyepiece—4.2 seconds down, 2.6 seconds up . . . 6.8 down, 4.0 up . . . 7.1 and 2.2 . . . 8.1 and 3.3.

I was starting to get the hang of it. But to do this right I needed to grab on to a single drop long enough to watch for the sudden variations in rise time, which would signal that it had gained or lost an electron. When I'd collected the data for a dozen drops and estimated their masses (with an equation called Stokes's law), I could calculate the fundamental unit of charge.

These things sound so easy in the physics books. You don't hear about the brass plates shorting out and sparking because a metal clip slipped into the wrong position. Or about spraying too much oil and clogging the pinhole. I'd confuse one drop with another or with a floater in my eye. I'd lock on to what seemed the perfect specimen and then watch helplessly as it drifted out of the focal plane. Sometimes a drop would be so heavy that it sank like a stone, or carry so much charge that when I turned on the voltage it rocketed out of sight. I tried and failed too many times before I realized: for me to master so delicate an experiment would be like learning to play the violin or at least make good cabinetry.

MAESTRO Millikan's touch was so deft that he could snag an oil drop in his gun sights, go home for dinner, and return later that evening to find it had barely moved. With his assistant Fletcher at his side, he'd call out the changes in speed as electrons hopped on and off a droplet like passengers riding a San Francisco cable car. If they needed a little boost, he opened a small lead door and zapped them with radium.

His data on the water drops had already come under attack from an Austrian experimenter who soon was claiming to have found "subelectrons" and suspected that there was no smallest unit of charge. But what Millikan had found with his earlier, cruder experiment was confirmed in spades by the oil drops. There really were electrons. One afternoon, Charles Proteus Steinmetz, the pioneering electrical engineer, came to watch the experiments. "I never would have believed it," he said, shaking Fletcher's hand. "I never would have believed it."

Early in 1910 they began writing up the results, and over the next three years Millikan continued to improve the experiment. The simple tabletop contraption morphed into a high-tech device with filtered air, tightly regulated temperature, pressure, and voltage, and a clock capable of marking time in milliseconds. Just as important was his progress in learning to read the drops. He recorded the ups and downs in his notebook:

> Very low something wrong . . . not sure of distance . . . Possibly a double drop . . . Beauty Publish . . . Good one for very small one . . . Exactly Right . . . Something the matter . . . Will not work out . . . Publish this Beautiful one.

As he tuned his reflexes the frequency of beauties increased:

> Perfect Publish . . . Best one yet.

It was as though the electrons themselves were shimmering in the light.

"*He who has seen that experiment . . . has in effect SEEN the electron,*" Millikan later wrote, italicizing his italics. "*He can count the number of electrons in a given small electrical charge with exactly as much certainty as he can attain in counting his fingers and his toes.*"

In 1913 he published his definitive value for the basic unit of electrical charge: 1.5924×10^{-19} coulomb. (The accepted value today is just slightly higher, $1.60217653 \times 10^{-19}$.) Ten years later he was awarded a Nobel Prize.

THE STORY has a strange denouement. After Millikan's former assistant, Harvey Fletcher, died in 1981, a memoir surfaced describing both his appreciation to Millikan for advancing his career and his disappointment at not getting more recognition for the oil-drop experiment. As Fletcher told the story, his professor showed up unexpectedly one day at his apartment offering to cut a deal. Millikan would be the sole author of the paper on the charge of the electron, but Fletcher would get full credit for a less important collaboration.

Fletcher's insistence that his account be published posthumously added to its credibility but also denied Millikan (who had died in 1953) an opportunity to respond. Judging from his autobiography, Millikan was not someone you'd want to be stuck with on a desert island, or even a cross-country flight. He could be patronizing and even a little bigoted. Though he was the indisputable force behind the isolation and measurement of the electron, he probably could have been more generous to his student. The beauty here lies with the experiment not the experimenter.

More troubling were accusations, coming still later, that

Millikan had cooked the books. The annotations in his laboratory journals, retrieved from the archives, were construed as evidence that he had combed his data for results that supported his preconceptions.

This is not an accusation that rings true to someone who has struggled with the oil-drop experiment. Millikan, I suspect, had simply developed a feeling for the mechanism, a sixth sense for when something had gone wrong: a slip of the thumb on the stopwatch, a sudden fluctuation in temperature or plate voltage, a dust particle masquerading as an oil drop. He knew when he had a bad run.

More interesting than the unfounded allegations is the question of how you keep from confusing your instincts with your suppositions, unconsciously nudging the apparatus, like a Ouija board, to come up with the hoped-for reply. It's something every experimenter must struggle with. The most temperamental piece of laboratory equipment will always be the human brain.

Afterword

The Eleventh Most Beautiful Experiment

IN THE AUTUMN of 2006, while I was science writer in residence at the Kavli Institute for Theoretical Physics in Santa Barbara, California, I gave a talk on *The Ten Most Beautiful Experiments.* Afterward a woman came forward to ask why there would be only men in the book.

I'd thought of including Marie Curie for her discovery of radium, laboriously distilling a smidgen of the glowing stuff from tons of radioactive ore. But that struck me as more of a heroic exploration than a controlled interrogation of nature. Lise Meitner seemed a likelier candidate, but her pioneering experiments in nuclear fission in the 1930s were done with Otto Hahn and Fritz Strassmann. Science was already becoming the collaborative effort that it is today. There were 439 names on the paper announcing the discovery of the top quark.

If I were to go beyond my arbitrary cutoff, maybe the eleventh most beautiful experiment would be Rita Levi-Montalcini's discovery of nerve-growth factor, Barbara McClintock's work on genetic regulation and jumping genes, or Chien-Shiung Wu's glorious demonstration that decaying electrons violate a law called conservation of parity.

I've barely finished the book and already I'm second-guessing myself. Why not Rutherford and the atomic nucleus, James Chadwick and the neutron, or Heike Kamerlingh Onnes and superconductivity? In biology there were Gregor Mendel with his garden experiments in genetics, and Oswald Avery, who showed that genes are made from DNA, a point beautifully driven home by Alfred Hershey and Martha Chase's famous Waring blender experiment. In what some have called the most beautiful experiment in biology Matthew Meselson and Franklin Stahl confirmed that DNA replicates as predicted by Watson and Crick's double helix.

As the twentieth century wears on, the pickings grow slimmer, with nature holding tightly to what secrets remain. The days when an unknown piece of the scaffolding could be exposed on a tabletop might be behind us. But you never know. The eleventh most beautiful experiment may be yet to come.

NOTES AND BIBLIOGRAPHY

THESE NOTES are intended to serve double duty as a suggested reading list, with the books grouped chapter by chapter. Because of the mercurial nature of the Web and the jagged appearance of URLs (never meant for human consumption) on the printed page, I've put links to Internet resources on my own site, talaya.net, where readers can find other supplementary material as well.

v "something like my own obituary": Paul Arthur Schilpp, *Albert Einstein: Philosopher-Scientist* (La Salle, Ill.: Open Court, 1979, originally published 1949), pp. 3, 9.

Prologue

xi "The appearance of this drop": Robert Millikan, *Physical Review* 32 (1911): 349, excerpted in Morris H. Shamos, *Great Experiments in Physics* (New York: Dover, 1987), p. 243.

xii The *Physics World* survey appeared in September 2002 (Robert P. Crease, "The Most Beautiful Experiment," pp. 19–20) and formed the basis of Crease's book *The Prism and the Pendulum: The Ten Most Beautiful Experiments in Science* (New York: Random House, 2003).

xiv "Questions of personal priority": Quoted in the first volume of Silvanus Phillips Thompson, *The Life of Lord Kelvin,* 2nd ed. (New York: Chelsea, 1976), p. 292.

1. *Galileo: The Way Things Really Move*

Drake, Stillman. *Galileo Studies: Personality, Tradition, and Revolution.* Ann Arbor: University of Michigan Press, 1970.

———. *Galileo at Work: His Scientific Biography.* Chicago: University of Chicago Press, 1978.

Galilei, Galileo. *Dialogues Concerning Two New Sciences.* Translated by Henry Crew and Alfonso de Salvio. Great Minds Series. Buffalo, N.Y.: Prometheus, 1991; originally published 1914.

———. *Two New Sciences, Including Centers of Gravity & Force of Percussion.* Translated by Stillman Drake. 2nd ed. New York: Modern Library, 2001; originally published 1974.

Koestler, Arthur. *The Sleepwalkers: A History of Man's Changing Vision of the Universe.* New York: Macmillan, 1959.

Rowland, Wade. *Galileo's Mistake: A New Look at the Epic Confrontation Between Galileo and the Church.* 1st U.S. ed. New York: Arcade, 2003.

Shea, William R., and Mariano Artigas. *Galileo in Rome: The Rise and Fall of a Troublesome Genius.* New York: Oxford University Press, 2003.

Sobel, Dava. *Galileo's Daughter: A Historical Memoir of Science, Faith, and Love.* New York: Walker, 1999.

3 epigraph: Galileo Galilei, *Discorsi e dimostrazioni matematiche intorno a due nuove scienze,* published in *Le Opere di Galileo Galilei,* edizione nazionale (Firenze: Tip. di G. Barbèra, 1890), p. 204. Translated into English by Henry Crew and Alfonso de Salvio as *Dialogues Concerning Two New Sciences* and by Stillman Drake as *Two New Sciences.* Quotations are from the Crew translation; pagination is from *Le Opere,* which Drake also uses.

4 debunking Galileo: Arthur Koestler is particularly harsh in *The Sleepwalkers,* pp. 425–509.

6 "Now you would not hide behind": *Opere,* p. 109.

6 "A piece of wooden moulding or scantling": Ibid., p. 213.

8 a little too good to be true: See Paul D. Sherman, "Galileo and the Inclined Plane Controversy," *Physics Teacher* 12 (1974): 343–48.

9 "A bronze ball rolling . . . !": Alexandre Koyré, "An Experiment in Measurement," *Proceedings of the American Philosophical Society* 97, no. 2 (1953): 222–37.

9 Stillman Drake on Galileo's inclined-plane experiment: "The Role of Music in Galileo's Experiments," *Scientific American* 232, no. 6 (June 1975): 98–104.

12 Galileo could have started with the odd-number progression: For more layers of Drake's analysis, see "Galileo's Discovery of the Law of Free Fall," *Scientific American* 228, no. 5 (May 1973): 84–92; the introduction to his translation of *Two New Sciences;* and his essay "Discovery of the Law of Fall," which is appended to the second edi-

tion. There is still more on the subject in his books *Galileo Studies,* pp. 214–39, and *Galileo at Work,* pp. 76–90.

13 Thomas Settle's reconstruction of the experiment: "An Experiment in the History of Science," *Science* 133 (1961): 19–23.

13 "The conductor of an orchestra": "Role of Music," p. 98.

15 "Even in his day": Ibid., p. 100.

2. *William Harvey: Mysteries of the Heart*

Aubrey, John, and Oliver Lawson Dick. *Aubrey's Brief Lives.* Ann Arbor: University of Michigan Press, 1957.

Harvey, William. *On the Motion of the Heart and Blood in Animals.* Translated by Robert Willis. Great Minds Series. Buffalo, N.Y.: Prometheus, 1993; originally published 1910.

———. *The Works of William Harvey.* Translated by Robert Willis. Classics in Medicine and Biology Series. Philadelphia: University of Pennsylvania Press, 1989; originally published 1965.

Keynes, G. L. *The Life of William Harvey.* New York: Oxford University Press, 1966.

Pagel, Walter. *William Harvey's Biological Ideas: Selected Aspects and Historical Background.* Basel, Switzerland: S. Karger, 1967.

———. *New Light on William Harvey.* Basel, Switzerland: S. Karger, 1983.

Park, Roswell. *An Epitome of the History of Medicine.* Philadelphia: F. A. Davis, 1897.

17 epigraph: Harvey, *On the Motion of the Heart and Blood in Animals.* (I have used Harvey's chapter and paragraph numbers.)

18 "Betwixt the visible and invisible": *Motion of the Heart,* IV.17.

18 "as though it had been seen through a window": Ibid., IV.16.

19 The best source of biographical details for Harvey is Keynes, *The Life of William Harvey.*

20 "He was wont to say": *Aubrey's Brief Lives,* pp. 130–31.

21 "For I could neither rightly perceive": *Motion of the Heart,* I.1.

23 "abundantly, impetuously": Ibid., IX.8.

24 "These two motions": Ibid., V.3–6.

25 "the sun of the microcosm": Ibid., VIII.3.

26 "Just as by air God makes ruddy": Stephen Finney Mason, *A History of the Sciences,* new rev. ed. (New York: Collier, 1962), p. 219.

27 Harvey decided to do the math: *Motion of the Heart,* IX.2–5.

27 "If a live snake be laid open": Ibid., X.6–7.

28 "detractors, mummers, and writers": "A Second Disquisition to John Riolan," *The Works of William Harvey,* p. 109.

30 "But he often sayd": *Aubrey's Brief Lives,* p. 128.

3. *Isaac Newton: What a Color Is*

Fauvel, John, ed. *Let Newton Be!* Reprint. New York: Oxford University Press, 1990.

Feingold, Mordechai. *The Newtonian Moment: Isaac Newton and the Making of Modern Culture.* New York: Oxford University Press, 2004.

Gleick, James. *Isaac Newton.* New York: Pantheon, 2003.

Hall, A. Rupert. *All Was Light: An Introduction to Newton's Opticks.* Reprint. New York: Oxford University Press, 1995.

Hooke, Robert. *Micrographia; or, Some Physiological Descriptions of Minute Bodies Made by Magnifying Glasses, with Observations and Inquiries Thereupon.* Dover Phoenix Editions, Mineola, N.Y.: Dover, 2003.

Sabra, A. I. *Theories of Light: From Descartes to Newton.* New York: Cambridge University Press, 1981.

Westfall, Richard S. *Never at Rest: A Biography of Isaac Newton.* New York: Cambridge University Press, 1980.

31 epigraph: From the preface of *Micrographia* (unpaginated).

34 For the early history of optics, see Sabra, *Theories of Light.*

34 Descartes describes his experiment with the water-filled sphere in his treatise "Les Méteores," which is extracted in William Francis Magie, *A Source Book in Physics* (New York, London: McGraw-Hill, 1935), pp. 273–78.

36 "Blue is an impression on the Retina": Hall, *All Was Light,* p. 18.

36 Newton's early optical experiments are described in his paper *Of Colours* (Cambridge University Library Add. Ms. 3975, pp. 1–22) and summarized and interpreted in *Isaac Newton,* pp. 79–89, *Never at Rest,* pp. 93–96, and *All Was Light,* pp. 33–38. All of Newton's scientific manuscripts—as well as his writings on alchemy and religion—are available online through the Newton Project.

36 "twixt your eye & a candle": *Of Colours,* 1. (I have used the paragraph numbers from the manuscript.)

37 Shut away from the plague: The precise chronology of Newton's optical experiments is somewhat confused, and there is reason to

wonder how much of the work was performed in Woolsthorpe and how much in Cambridge. See *Never at Rest,* pp. 156–58.

37 "multitude of reflecting surface": *Of Colours,* 56.

37 "Accordingly as the glasses": Ibid., 36.

38 "betwixt my eye & the bone": Ibid., 58–60.

38 "from the center greene": Ibid., 62–63.

38 "a vast multitud of these slender pipes": Ibid., 64.

40 "good deepe red": Ibid., 6.

40 The experiment with the window shutter and prism is described in "Fair Copy of 'A Theory Concerning Light and Colors' " (Cambridge University Library, Add. Ms. 3970.3ff.), pp. 460–66. It was later published as "New Theory About Light and Colors," *Philosophical Transactions of the Royal Society* 80 (19 February 1671–1672). Both versions are online at the Newton Project. For narratives and analyses of the experiment, see *Never at Rest,* pp. 156–75, and *Theories of Light,* pp. 234–44.

40 "It was at first a very pleasing divertisement": All quotations here are from "Theory Concerning Light and Colors."

41 a multitude of experiments: Westfall lays out the details in *Never at Rest,* pp. 94–96, and notes that Newton had a hunch about the heterogeneity of white light as early as 1664 in Cambridge.

42 "blew rays suffer": *Of Colours,* 6.

42 "consists of rayes differently refrangible": "Theory Concerning Light and Colors."

42 "To the same degree of refrangibility": Ibid.

44 "bright cloud": "A Letter of the Learn'd Franc. Linus . . . animadverting upon Mr Newtons Theory of Light and Colors" and "An Answer to this Letter," *Philosophical Transactions of the Royal Society* 110 (25 January 1674–1675). Available online at the Newton Project.

4. *Antoine-Laurent Lavoisier: The Farmer's Daughter*

Bell, Madison Smartt. *Lavoisier in the Year One: The Birth of a New Science in an Age of Revolution.* New York: Norton, 2005.

Djerassi, Carl, and Roald Hoffmann. *Oxygen: A Play in Two Acts.* New York: Wiley-VCH, 2001.

Donovan, Arthur. *Antoine Lavoisier: Science, Administration and Revolution.* New ed. New York: Cambridge University Press, 1996.

Guerlac, Henry. *Antoine-Laurent Lavoisier, Chemist and Revolutionary.* New York: Scribner, 1975.

———. *Lavoisier—The Crucial Year: The Background and Origin of His First Experiments on Combustion in 1772.* Ithaca, N.Y.: Cornell University Press, 1961.

Holmes, Frederic Lawrence. *Antoine Lavoisier: The Next Crucial Year; or, The Sources of His Quantitative Method in Chemistry.* Princeton, N.J.: Princeton University Press, 1998.

Lavoisier, Antoine-Laurent. *Elements of Chemistry.* New York: Dover, 1965.

Poirier, Jean-Pierre. *Lavoisier: Chemist, Biologist, Economist.* Reprint. Philadelphia: University of Pennsylvania Press, 1998.

45 epigraph: Djerassi and Hoffmann, *Oxygen,* p. 119.

46 the particle accelerator of its day: Donovan, *Antoine Lavoisier,* p. 47.

46 The diamond-burning experiment is described in Poirier, *Lavoisier,* pp. 58–60.

47 "the air contained in matter": Poirier, *Lavoisier,* p. 58.

47 "Camphire dissolved in well deflegmed spirit": Portsmouth Collection (Add. Ms. 3975), Cambridge University Library, Cambridge University, pp. 32–44.

48 "In [Saturn] is hid an immortal soul": The manuscript is at Yale University (Beinecke Library, Mellon Ms. 79) and online at the Newton Project. The passage was copied from George Starkey's *The Marrow of Alchemy* (1654). For the meaning of the alchemical terms, I relied on an analysis by William Newman, a historian of science at Indiana University, on the PBS Web site for the *Nova* show "Newton's Dark Secrets."

48 For the history of the phlogiston hypothesis, see Stephen Finney Mason, *A History of the Sciences,* new rev. ed. (New York: Collier, 1962), pp. 303–13.

49 "impelled by forces": Poirier, *Lavoisier,* p. 63.

49 "gave wings to earthly molecules": Ibid., p. 62.

50 Lavoisier's 1769 experiment: Ibid., pp. 32–34.

52 Lavoisier's marriage: Ibid., pp. 39–41.

52 Details about Lavoisier's wife and her role in his experiments are in Roald Hoffmann, "Mme. Lavoisier," *American Scientist* 90, no. 1 (January–February 2002): 22. A virtual museum of Lavoisier's work, including a detailed chronology of his experiments and photographs of some of his equipment, is being assembled online at Panopticon Lavoisier. In addition, the complete works of Lavoisier, in French, are available on the Web at *Les Œuvres de Lavoisier.*

52 "different kinds of air": Priestley wrote a three-volume work, *Exper-*

iments and Observations on Different Kinds of Air (London: printed for J. Johnson, 1774). For a brief history of this work, see Mason's *History of the Sciences*, pp. 304–6.

53 Lavoisier's experiments with phosphorus, sulfur, tin, and litharge: Poirier, *Lavoisier*, pp. 65–66, and Guerlac, *Lavoisier*, pp. 79–80. The tin experiment is described in *History of the Sciences*, p. 308. The apparatus in the litharge experiment, called a pneumatic trough, was a variation of one devised by Stephen Hales.

53 He thought he knew the answer: In the experiments with phosphorus and sulfur he also saw signs of air absorption, and a Parisian chemist had reported a similar result. See Guerlac, *Lavoisier*, p. 79.

54 Estimates of the price of *mercurius calcinatus* are from Poirier, *Lavoisier*, p. 74.

54 "without addition": Ibid.

54 "What surprised me more": Ibid.

54 "I fancied that my breast": Poirier, *Lavoisier*, p. 76.

55 Lavoisier's first experiments with mercury are described in Poirier, *Lavoisier*, pp. 79–80.

55 "eminently breathable": Ibid., p. 103. Lavoisier used these words in a talk to the French academy in April 1775 that was published as "On the Nature of the Principle Which Combines with Metals During Calcinations and Increases Their Weight." Three years later, he revised the paper with his new interpretation. James Bryant Conant compared the two versions in "The Overthrow of the Phlogiston Theory," in *Harvard Case Histories in Experimental Science*, vol. 1 (Cambridge, Mass.: Harvard University Press, 1957).

55 Lavoisier reported the results of his experiment with the matrass to the Académie des Sciences on May 3, 1777, as "Experiments on the Respiration of Animals and on the Changes Which Happen to Air in Its Passage Through Their Lungs," and later in chapter 3 of his *Elements of Chemistry*, pp. 32–37.

57 "when a taper was plunged": *Elements of Chemistry*, p. 35.

57 "with a dazzling splendor": Ibid., p. 36.

57 "Here is the most complete kind of proof": Poirier, *Lavoisier*, p. 104.

58 Lavoisier's execution: Ibid., pp. 381–82.

59 A story ricocheting through the Internet: It apparently originated with a comment made on a Discovery Channel program, and in some versions the assistant counting the blinks was Lagrange. For a debunking of the legend, see William B. Jensen, "Did Lavoisier Blink?" *Journal of Chemical Education* 81 (2004): 629.

5. *Luigi Galvani: Animal Electricity*

Fara, Patricia. *An Entertainment for Angels: Electricity in the Enlightenment.* New York: Columbia University Press, 2003.

Galvani, Luigi. *Galvani Commentary of the Effect of Electricity and Muscular Motion.* Translated by Robert Montraville Green. Cambridge, Mass.: E. Licht, 1953.

Heilbron, J. L. *Electricity in the 17th and 18th Centuries: A Study of Early Modern Physics.* Berkeley: University of California Press, 1979.

Ostwald, Wilhelm. *Electrochemistry: History and Theory.* New Delhi: Amerind. Published for the Smithsonian Institution and the National Science Foundation, Washington, D.C., 1980.

Pancaldi, Giuliano. *Volta: Science and Culture in the Age of Enlightenment.* Princeton, N.J.: Princeton University Press, 2005.

Pera, Marcello. *The Ambiguous Frog: The Galvani-Volta Controversy on Animal Electricity.* Princeton, N.J.: Princeton University Press, 1992.

60 epigraph: Galvani, *De Viribus Electricitatis in Motu Musculari Commentarius,* p. 40. (Unless otherwise indicated all quotations are from the English translation by Robert Montraville Green, *Galvani Commentary on the Effect of Electricity and Muscular Motion.*)

61 Symmer's experiment is described in Heilbron, *Electricity in the 17th and 18th Centuries,* pp. 431–37, and in Pera, *The Ambiguous Frog,* pp. 38–39.

61 "When this experiment is performed": *The Ambiguous Frog,* p. 39, quoting from Robert Symmer, "New Experiments and Observations Concerning Electricity," *Philosophical Transactions* 61 (1759): 340–89.

61 The eighteenth-century electrical vogue is described in *Electricity in the 17th and 18th Centuries,* pp. 263–70; *The Ambiguous Frog,* pp. 3–18; and Fara, *An Entertainment for Angels.*

63 "said to proceed from some animals": *The Ambiguous Frog,* pp. 60–61, quoting Priestley's *Experiments and Observations on Different Kinds of Air,* pp. 277–79.

64 Galvani's experiment near the Palazzo Zamboni is described in *The Ambiguous Frog,* p. 80.

67 The experiments with the railing and silver box: *Commentary,* pp. 40–41, and *The Ambiguous Frog,* pp. 81–83.

67 "crept into the animal and accumulated": *Commentary,* p. 40.

67 "At the very moment the foot touched": *The Ambiguous Frog,* p. 82. Galvani describes the scene in *Commentary,* pp. 43–44.

68 Galvani's wide-ranging speculations are in *Commentary*, pp. 78–81.
68 "But let there be a limit to conjectures!": Ibid., p. 81.
68 "among the demonstrated truths": *The Ambiguous Frog*, p. 98.
68 "the same convulsions, spasms and jerks": Ibid., p. 100.
69 Volta's experiment with tin and silver clips: Ibid., p. 105.
69 "Galvani's theory and explanations": Ibid., p. 114.
69 "If that is how things are": Ibid., p. 113.
70 the Galvanists' experiments challenging Volta's bimetallic hypothesis: Ibid., pp. 119–22.
70 "Why then ascribe": Ibid., p. 122.
71 "Each time I touch it": Ibid., p. 123.
71 Galvani's experiment without external conductors (often referred to as his "third experiment"): Ibid., p. 129.
71 "But if that is how things are": *The Ambiguous Frog*, p. 13.
72 Volta's battery: described in ibid., pp. 153–58.
73 Galvani's final ("fourth") experiment: Ibid., pp. 147–48.
74 "Now what dissimilarity": Ibid., p. 148.

6. *Michael Faraday: Something Deeply Hidden*

Cantor, Geoffrey. *Michael Faraday, Sandemanian and Scientist.* New ed. London: Palgrave Macmillan, 1993.

Dibner, Bern. *Oersted and the Discovery of Electromagnetism.* Norwalk, Conn.: Burndy Library, 1961.

Faraday, Michael. *The Chemical History of a Candle.* New York: Dover, 2003; originally published 1861.

———. *Experimental Researches in Electricity.* New York: Dover, 1965; originally published 1839–1855.

———. *The Forces of Matter.* Great Minds Series. Buffalo, N.Y.: Prometheus, 1993.

Faraday, Michael, and Howard J. Fisher. *Faraday's Experimental Researches in Electricity: Guide to a First Reading.* Santa Fe, N.M.: Green Lion, 2001.

Faraday, Michael, and Thomas Martin. *Faraday's Diary.* London: Bell, 1932.

Hamilton, James. *A Life of Discovery: Michael Faraday, Giant of the Scientific Revolution.* New York: Random House, 2004.

Jones, Bence. *The Life and Letters of Faraday.* London: Longmans, Green, 1870.

Lehrs, Ernst. *Spiritual Science: Electricity and Michael Faraday.* London: Rudolph Steiner Press, 1975.

Russell, Colin Archibald. *Michael Faraday: Physics and Faith.* New York: Oxford University Press, 2000.

Williams, L. Pearce. *Michael Faraday: A Biography.* New York: Da Capo, 1987.

Woolley, Benjamin. *The Bride of Science: Romance, Reason, and Byron's Daughter.* New York: McGraw-Hill, 2000.

75 first epigraph: Jones, *The Life and Letters of Faraday,* vol. 2, pp. 473–74. Jones dates the letter April 22, 1867.

75 second epigraph: Faraday, *Experimental Researches in Electricity,* Third Series, para. 280.

76 "Enchantress of Numbers": Woolley, *The Bride of Science,* p. 274.

76 "Bride of Science": Ibid., p. 306.

76 "calculus of the nervous system": Ibid., p. 305.

76 "ladye-fairy": Hamilton, *A Life of Discovery,* p. 318.

78 Oersted described his discovery in "Experiments on the Effect of a Current of Electricity on the Magnetic Needle," *Annals of Philosophy* 16 (1820): 276.

79 Faraday's experiments with a crude electric motor are described in *Faraday's Diary,* pp. 50–51, and are summarized in Williams, *Michael Faraday,* p. 156, and *A Life of Discovery,* pp. 164–65.

79 errands of the Industrial Age: *A Life of Discovery,* pp. 151–56.

79 "unfortunately occupied": Williams, *Michael Faraday,* p. 109.

79 crispations: Ibid., pp. 177–78.

80 "Mercury on tin plate": *A Life of Discovery,* pp. 236–37, quoting Faraday's diaries.

80 The induction ring experiment is described in Williams, *Michael Faraday,* pp. 182–83; in *Faraday's Diary,* August 29, 1831, p. 367; and in *Experimental Researches in Electricity,* First Series, para. 27–28.

81 "wave of electricity": Williams, *Michael Faraday,* p. 183.

82 as a German scientist proposed: This was Johann Ritter, ibid., pp. 228–30.

82 Faraday's breakdown: *A Life of Discovery,* pp. 293–94.

82 "You drive me to desperation": Ibid., p. 319.

82 Maybe it is too great a reach: Another inspiration for undertaking the polarization experiment may have been a letter from William Thomson, the future Lord Kelvin: Williams, *Michael Faraday,* pp. 383–84.

83 a question that had been gnawing at him: Faraday describes an ear-

lier attempt using a trough of electrolytes in his diary entry for September 10, 1822, p. 71.

83 "so that they might regulate": From the official history posted on the Trinity House Web site.

83 Faraday's work in lighthouses: *A Life of Discovery,* pp. 322–23.

83 Faraday describes his experiments with light beams in vol. 4 of his diary, paragraphs 256–67, and in the Nineteenth Series of *Experimental Researches,* 2146–72. There is also an account in Williams, *Michael Faraday,* pp. 384–87.

85 "At present I have scarcely a moment": *A Life of Discovery,* p. 327.

86 "BUT when contrary magnetic poles": Williams, *Michael Faraday,* p. 386. The 1932 edition of the diary, edited by Thomas Martin, includes a facsimile of the handwritten page with the triple underscoring.

86 "ALL THIS IS A DREAM": *A Life of Discovery,* p. 334.

86 "You see what you do": Ibid., p. 320.

7. James Joule: How the World Works

Baeyer, Hans Christian von. *Maxwell's Demon: Why Warmth Disperses and Time Passes.* New York: Random House, 1999.

Caneva, Kenneth L. *Robert Mayer and the Conservation of Energy.* Princeton, N.J.: Princeton University Press, 1993.

Cardwell, Donald S. L. *From Watt to Clausius: The Rise of Thermodynamics in the Early Industrial Age.* London: Heinemann, 1971.

———. *James Joule: A Biography.* Manchester, England: Manchester University Press, 1991.

———. *Wheels, Clocks, and Rockets: A History of Technology.* New York: Norton, 2001.

Carnot, Sadi. *Reflections on the Motive Power of Fire: And Other Papers on the Second Law of Thermodynamics.* New York: Dover, 2005.

Joule, James Prescott, William Scoresby, Lyon Playfair, and William Thomson. *The Scientific Papers of James Prescott Joule.* London: The Society, 1963; originally published 1887.

Lindley, David. *Degrees Kelvin: A Tale of Genius, Invention, and Tragedy.* Washington, D.C.: Joseph Henry Press, 2005.

Thompson, Silvanus Phillips. *The Life of Lord Kelvin.* 2nd ed. New York: Chelsea, 1977; originally published 1910.

Truesdell, Clifford A. *The Tragicomical History of Thermodynamics, 1822–1854.* New York: Springer, 1980.

88 epigraph: Truesdell, *The Tragicomical History of Thermodynamics,* pp. 164–65.

89 "inoculated with Faraday fire": Thompson, *The Life of Lord Kelvin,* p. 19.

89 The encounter on the trail with Kelvin is described in *Life of Lord Kelvin,* p. 265, and in Cardwell, *James Joule,* pp. 88–89.

90 Joule and Thomson's meeting in Oxford: *James Joule,* pp. 82–83, and Lindley, *Degrees Kelvin,* pp. 74–75.

91 "Joule is, I am sure": *James Joule,* p. 85.

92 The marriage of Rumford to Mme. Lavoisier is recounted in Poirier, *Lavoisier,* pp. 407–9 (cited in my notes for chapter 4). Poirier also describes, pp. 125–26, an extramarital affair she had with the economist Pierre Samuel du Pont de Nemours, father of the founder of the chemical company.

92 "en bon point": Ibid., p. 407.

92 "merely by the strength": Benjamin Thompson, "An Inquiry Concerning the Source of the Heat Which Is Excited by Friction," *Philosophical Transactions of the Royal Society* 88 (1798): 80–102; excerpted in Magie's *A Source Book in Physics* (cited in my notes for chapter 3), pp. 159–60.

93 "a very *brisk* and *vehement agitation*": Hooke, *Micrographia.* Observ. VI. "Of Small Glass Canes" (cited in my notes for chapter 3), p. 12.

96 Joule's shocking childhood experiments and other biographical details are from *James Joule,* pp. 13–16.

96 "I can hardly doubt that electro-magnetism": *The Scientific Papers of James Prescott Joule,* vol. 1, p. 14.

97 "there seemed to be nothing to prevent": Ibid., p. 47; *James Joule,* p. 36.

97 Joule's motors are described in *Scientific Papers,* vol. 1, pp. 1–3, 16–17, and in *James Joule,* pp. 32–37.

98 "The comparison is so very unfavourable": *James Joule,* p. 37, quoting a public lecture at the Royal Victoria Gallery, February 16, 1841.

98 Joule reported on his experiment with the crank in part 1 of Joule, "On the Calorific Effects of Magneto-Electricity, and on the Mechanical Value of Heat," *Scientific Papers,* vol. 1, pp. 123–59; see also *James Joule,* pp. 53–56.

100 The experiment with the pulleys is in part 2 of "Calorific Effects," pp. 149–57, and in *James Joule,* pp. 56–58.

100 "the subject did not excite": *Scientific Papers,* vol. 2, p. 215.

101 Joule published the experiment presented at Oxford as "On the Mechanical Equivalent of Heat, as Determined by the Heat Evolved by the Friction of Fluids," *Scientific Papers,* vol. 2, pp. 277–81. For later refinements see the similarly titled "On the Mechanical Equivalent of Heat," *Scientific Papers,* vol. 1, pp. 298–328.

103 "irrecoverably lost": *Life of Lord Kelvin,* p. 288.

103 "Within a finite period of time past": Ibid., p. 291.

8. *A. A. Michelson: Lost in Space*

Livingston, Dorothy Michelson. *Master of Light: A Biography of Albert A. Michelson.* Reprint. Chicago: University of Chicago Press, 1979.

Mach, Ernst. *The Principles of Physical Optics: An Historical and Philosophical Treatment.* Translated by John S. Anderson and A. F. A. Young. London: Methuen, 1926; originally published 1921.

Maxwell, James Clerk. *Matter and Motion.* New York: Dover, 1952; originally published 1876.

Michelson, Albert Abraham. *Experimental Determination of the Velocity of Light.* Minneapolis: Lund, 1964. A reproduction of Michelson's handwritten report on his experiments of 1878, commissioned by Honeywell, Inc.

———. *Light Waves and Their Uses.* Chicago: University of Chicago Press, 1961; originally published 1903.

———. *Studies in Optics.* Phoenix Science Series. Chicago: University of Chicago Press, 1962; originally published 1927.

Swenson, Lloyd S. *Ethereal Aether: A History of the Michelson-Morley-Miller Aether-Drift Experiments, 1880–1930.* Austin: University of Texas Press, 1972.

104 epigraph: Maxwell, *Matter and Motion,* quoted in Swenson, *Ethereal Aether,* p. 30.

105 Michelson's breakdown: Livingston, *Master of Light,* pp. 111–15.

105 "soft in the head": In a letter to his father, September 27, 1885, Morley referred to "some symptoms which point to softening of the brain." Quoted in *Master of Light,* p. 112.

105 "rendering all the fancies, moods, and emotions": Michelson, *Light Waves and Their Uses*, p. 2.

106 Galileo's experiment on the speed of light: *Two New Sciences* (cited in the notes for chapter 1), *Opere*, p. 88.

106 "if not instantaneous": Ibid.

106 early history of light-speed measurements: *Master of Light*, pp. 47–49, and Norriss S. Hetherington, "Speed of Light," in J. L. Heilbron, ed., *The Oxford Companion to the History of Modern Science* (New York: Oxford University Press, 2003), pp. 467–68.

106 Roemer's paper on the speed of light was translated into English as "A Demonstration Concerning the Motion of Light," *Philosophical Transactions of the Royal Society* 12 (June 25, 1677): 893–94. Bradley described stellar aberration in "An Account of a New Discovered Motion of the Fixed Stars," *Philosophical Transactions of the Royal Society* 35 (1727–28): 637–61. Both can be found in Magie's *A Source Book in Physics* (cited in the notes for chapter 3), pp. 335–40. The actual values of their estimates vary depending on whether they are based on what was known then or now about planetary distances. I used the numbers in the *Encyclopaedia Britannica* entries for Roemer and Bradley.

107 Fizeau's experiment appeared as "Sur un expérience relative à la vitesse de propagation de la lumiére," *Comptes Rendus* 29 (1849): 90. An English translation is in *Source Book in Physics*, pp. 341–42.

108 Foucault described his experiment in "Détermination expérimentale de la vitesse de la lumière: parallaxe du Soleil," *Comptes Rendus* 55 (1862): 501–3, 792–96, which is excerpted in *Source Book in Physics*, pp. 343–44.

109 Michelson's early biographical history is from *Ethereal Aether*, pp. 33–43, and *Master of Light*, pp. 11–44.

110 Michelson describes his light-speed experiment in "Experimental Determination of the Velocity of Light," *Proceedings AAAS*, vol. 27 (1878), pp. 71–77. For a summary see *Master of Light*, pp. 51–63. His original handwritten paper was reprinted and published as a facsimile by Honeywell in 1964.

110 "being about 200 times that": *Velocity of Light*, p. 5.

111 "It would seem that the scientific world": *Master of Light*, p. 63.

111 "globular bodyes" and "a Tennis-ball struck with an oblique Racket": Newton used these words in "A Theory Concerning Light and Colors," cited in the notes for chapter 3.

111 "fits of easy reflexion": Newton uses the term in his *Opticks; or, A*

Treatise of the Reflections, Refractions, Inflexions and Colours of Light, 2nd ed., with additions (London: 1717), 3rd book, part 1, p. 323.

112 The trip to Europe is described in *Ethereal Aether,* pp. 67–68. According to *Master of Light,* pp. 74–75, Michelson goes to Berlin first and then to Paris in 1881.

113 "struggling upstream and back": *Master of Light,* p. 77.

113 Michelson's Berlin and Potsdam experiments: Michelson, "The Relative Motion of the Earth and the Luminiferous Aether," *American Journal of Science,* Third Series, 22 (August): 120–29. Described in *Ethereal Aether,* pp. 68–73, and *Master of Light,* pp. 77–84.

113 "So extraordinarily sensitive": Albert A. Michelson and Edward W. Morley, "On the Relative Motion of the Earth and the Luminiferous Ether," *American Journal of Science,* Third Series, vol. 34, no. 203 (November 1887), p. 124. The event he refers to took place in Potsdam.

115 "I have a very high respect": Bell made the observation in 1883 in a letter to his wife quoted in *Master of Light,* pp. 96–97.

116 measuring light speed in a vacuum: Ibid., pp. 95–96.

117 repeating the Fizeau experiment: *Ethereal Aether,* pp. 81–87, and *Master of Light,* pp. 110–11.

117 the fire at Case: *Master of Light,* pp. 121–22.

119 "if light travels with the same velocity": Morely wrote this in a letter of April 17, 1887, to his father; quoted in *Ethereal Aether,* p. 91.

119 the Michelson-Morley experiment: "The Relative Motion," summarized in *Ethereal Aether,* pp. 91–97, and *Master of Light,* pp. 126–33.

120 Miller on Mount Wilson: *Ethereal Aether,* pp. 205–6.

120 Michelson on Mount Wilson: Ibid., pp. 225–26.

120 "one of the grandest generalizations": *Light Waves and Their Uses,* p. 162.

120 It took the publication of Einstein's special theory: Einstein, however, denied that the Michelson-Morley results were in themselves a motivation for his work.

9. *Ivan Pavlov: Measuring the Immeasurable*

Babkin, B. P. *Pavlov.* Chicago: University of Chicago Press, 1975.

Frolov, Y. P. *Pavlov and His School: The Theory of Conditioned Reflexes.* New York: Johnson Reprint, 1970.

Gray, Jeffrey A. *Ivan Pavlov.* New York: Viking, 1980.

James, William. *The Principles of Psychology.* New York: Dover, 1950; originally published 1890.

Pavlov, Ivan Petrovich. *Conditioned Reflexes: An Investigation of the Physiological Activity of the Cerebral Cortex.* Translated by G. V. Anrep. New York: Dover, 1960; originally published 1927.

———. *Lectures on Conditioned Reflexes.* Vol. 1. Translated by W. Horsley Gantt. New York: International, 1928; originally published 1923.

Sechenov, Ivan. *Reflexes of the Brain.* Cambridge, Mass.: MIT Press, 1965.

Todes, Daniel Philip. *Ivan Pavlov: Exploring the Animal Machine.* New York: Oxford University Press, 2000.

———. *Pavlov's Physiology Factory: Experiment, Interpretation, Laboratory Enterprise.* Baltimore: Johns Hopkins University Press, 2002.

121 epigraph: Todes, *Pavlov's Physiology Factory,* p. 123, quoting Pavlov's 1893 essay, "Vivisection."

122 The names of some of Pavlov's dogs are in Tim Tully, "Pavlov's Dogs" (*Current Biology* 13, no. 4: R117–19), and on a page at the Web site for Cold Spring Harbor Laboratory.

123 "When I dissect and destroy": Babkin, *Pavlov,* p. 162.

123 Details of Pavlov's life are from a biographical sketch by his colleague and translator W. Horsley Gantt in Pavlov, *Lectures on Conditioned Reflexes,* pp. 11–31; *Pavlov,* pp. 5–23; and Todes, *Ivan Pavlov,* pp. 11–43.

123 Pavlov and the library: Todes, *Ivan Pavlov,* p. 19.

124 "Absolutely all the properties": Sechenov, *Reflexes of the Brain,* p. 4.

125 "complex chemical factory": Todes, *Ivan Pavlov,* p. 59.

125 Pavlov's digestive experiments: Todes, *Ivan Pavlov,* pp. 53–65; *Pavlov,* pp. 224–30; and Gray, *Ivan Pavlov,* pp. 20–25.

126 "Every material system": Pavlov, *Conditioned Reflexes,* lecture 1, p. 8.

127 an honor he was almost denied: For a fascinating account of the Nobel politicking, see *Pavlov's Physiology Factory,* pp. 332–45.

127 "It is clear that we did not": *Pavlov,* p. 229.

127 Pavlov's firsthand accounts of his salivation experiments can be found in his two books, *Conditioned Reflexes* and *Lectures on Conditioned Reflexes,* vol. 1. Good secondary sources are Gray, *Ivan Pavlov,* pp. 26–51, and Todes, *Ivan Pavlov,* pp. 71–79.

127 "convinced of the uselessness": *Lectures on Conditioned Reflexes,* p. 71.

128 "Indeed, what means have we": *Pavlov,* p. 277.

128 "Does not the eternal sorrow": *Lectures on Conditioned Reflexes,* p. 50.

128 "But now the physiologist turns": Ibid., p. 121.

129 "The self-same atoms": James, *The Principles of Psychology,* p. 146.

129 "primordial mind-dust": Ibid., p. 150.
129 "Just as the material atoms": Ibid., p. 149.
130 "The soul stands related": Ibid., p. 131.
130 Benjamin Libet describes his experiments on free will in *Mind Time: The Temporal Factor in Consciousness* (Cambridge, Mass.: Harvard University Press, 2004).
130 "If we knew thoroughly": *Principles of Psychology*, pp. 132–33.
131 "The naturalist must consider": *Lectures on Conditioned Reflexes*, p. 82.
132 drooling at an electric shock and after a three-minute delay: Ibid., pp. 149, 186–87.
132 salivating on the half hour: *Conditioned Reflexes*, lecture 3, p. 41.
132 "I am convinced": *Lectures on Conditioned Reflexes*, p. 233.
132 discriminating between clockwise and counterclockwise, etc.: *Conditioned Reflexes*, lecture 7, pp. 117–30, and lecture 13, p. 222; *Lectures on Conditioned Reflexes*, p. 140.
133 "Footfalls of a passer-by": *Conditioned Reflexes*, lecture 2, p. 20.
133 "Tower of Silence": *Lectures on Conditioned Reflexes*, pp. 144–46; Frolov, *Pavlov and His School*, pp. 60–62; and Todes, *Ivan Pavlov*, pp. 77–78.
133 "a submarine ready for battle": *Pavlov and His School*, p. 61.
133 Pavlov describes the experiment on ascending and descending scales in *Lectures on Conditioned Reflexes*, p. 141. (The notes were D, E, F-sharp, and G-sharp.)
134 "The movement of plants": Ibid., p. 59.
135 conditional reflexes: Todes and others argue that this is a better translation of Pavlov's term *uslovnyi refleks* than the more familiar "conditioned reflex." See *Pavlov's Physiology Factory*, pp. 244–46.
135 A copy was tracked down: Tim Tully described his search in "Pavlov's Dogs" (*Current Biology* 13, no. 4: R118).
136 "Pavlov's flies": from a Cold Spring Harbor press release, February 17, 2003, available on the lab's Web site.
137 "Let the dog, man's helper": Todes, *Ivan Pavlov*, p. 100.

10. *Robert Millikan: In the Borderland*

Goodstein, Judith R. *Millikan's School: A History of the California Institute of Technology.* New York: Norton, 1991.
Holton, Gerald James. *The Scientific Imagination: Case Studies.* New York: Cambridge University Press, 1978.

Millikan, Robert Andrews. *Autobiography.* London: Macdonald, 1951.
———. *The Electron: Its Isolation and Measurement and the Determination of Some of Its Properties.* Chicago: University of Chicago Press, 1924.
———. *Evolution in Science and Religion.* New Haven: Yale University Press, 1927.
———. *Science and Life.* Boston: Pilgrim, 1924.
Thomson, Joseph John. *Recollections and Reflections.* New York: Macmillan, 1937.
Weinberg, Steven. *The Discovery of Subatomic Particles.* New York: Freeman, 1990.

138 epigraph: William Crookes, "On Radiant Matter II," *Nature* 20 (September 4, 1879): 439–40.

140 Millikan misremembered the talk as occurring on Christmas Eve: He described listening to the Roentgen lecture in *Evolution in Science and Religion,* pp. 10–11. The actual date of the meeting was January 4, 1896. (Roentgen had, in fact, given another talk in December 1895 in Würzburg.) The oft-cited Michelson lecture is recalled in *Autobiography,* pp. 39–40.

140 Hertz on radio waves and light: "On Electric Radiation," *Annalen der Physik* 36 (1889): 769; in *A Source Book in Physics,* pp. 549–61.

141 "We had not come quite as near": *Autobiography,* p. 11.

141 The history of the discharge-tube experiments is described in Weinberg, *The Discovery of Subatomic Particles,* pp. 20–25, 102–5.

141 Crookes's work is described in two beautifully illustrated papers in volume 20 of *Nature:* "On Radiant Matter" (August 28, 1879): 419–23, and "On Radiant Matter II" (September 4, 1879): 436–40. Both are reprinted in David M. Knight, *Classical Scientific Papers: Chemistry, Second Series. Papers on the Nature and Arrangement of the Chemical Elements* (New York: American Elsevier, 1970), pp. 89–98.

143 "A fourth state of matter": "Radiant Matter II," 439. Crookes borrowed the term from Faraday.

143 Roentgen's penetrating rays: "On a New Kind of Rays," translated by Arthur Stanton, *Nature* 53 (1896): 274–76. A different translation is excerpted in *A Source Book in Physics* (cited in my notes for chapter 3), pp. 600–10.

143 Becquerel's uranium experiment: "On the Rays Emitted by Phosphorescence," *Comptes Rendus* 122 (1896): 420–21, 501–3; in *A Source Book in Physics,* pp. 610–13.

143 J. J. Thomson described his experiments in "Cathode Rays," *Philosophical Magazine* 44, no. 293 (1897): 293–316. A facsimile appears in Stephen Wright, *Classical Scientific Papers: Physics* (New York: American Elsevier, 1964). Weinberg analyzes the experiment in *Discovery of Subatomic Particles*, pp. 12–71.

143 Electrons: The name had first been used by the Irish physicist George Johnstone Stoney in "Of the Electron or Atom of Electricity," *Philosophical Magazine* 38 (1894), p. 418.

144 My Thomson apparatus, made by Leybold, also included a focusing grid, or Wehnelt (named after the German physicist who invented it).

146 2.5×10^8 coulombs of charge per gram: The formula for the charge-to-mass ratio is v/Br where v is the velocity of the electrons, B is the strength of the magnetic field, and r is the radius of the curving beam. This turns out to be equivalent to

$$\frac{2V\,(5/4)^3 a^2}{(N\,\mu_0 I r)^2}$$

a = the radius of the coils
N = the number of turns of wire in the coils
V = the accelerating voltage on the anode
I = the amperes of current in the coils
r = the radius of beam

μ_0 is a number called the permeability constant ($4\pi \times 10^{-7}$), a conversion factor that makes all the units—volts, amperes, coulombs, centimeters, and grams—play well together.

146 the quantity of electricity flowing each second through a 100-watt bulb: assuming, of course, a power source of 100 volts.

146 The value for the electron was about a thousand times greater: Thomson also considered the possibility that electrons might have a larger mass and a smaller charge, but that would have contradicted experiments by Philip Lenard suggesting that cathode-ray corpuscles were considerably lighter than molecules of air.

146 feeling like a has-been: Millikan tells the story in *Autobiography*, pp. 84–85.

147 The Cavendish experiment with the vapor cloud was done by H. A. Wilson and marked an improvement over earlier attempts by Thomson and J. S. E. Townsend. The experiments are summarized in *Autobiography*, pp. 85–87, and in *The Electron*, pp. 45–57. Weinberg analyzes the work in *Discovery of Subatomic Particles*, pp. 91–95. The device used in the experiments, a Wilson cloud chamber, was

invented by the Scotsman C. T. R. Wilson, who used it to observe the tracks of cosmic rays.

147 "like Mohammed's coffin": *Autobiography,* p. 89. Curiously, Thomson used the same analogy fourteen years earlier in his own memoirs, *Recollections and Reflections,* p. 343.

148 Millikan's water-drop experiments: *Autobiography,* pp. 89–91. For an analysis see "Subelectrons, Presuppositions and the Millikan-Ehrenhaft Dispute," in Holton, *The Scientific Imagination,* pp. 42–46.

148 "1, 2, 3, 4, or some other exact multiple": *Autobiography,* p. 90. Millikan reported his result as 4.65×10^{-10} electrostatic units (also called statcoulombs), which converts to 1.55×10^{-19} coulombs.

149 The Winnipeg meeting is described in *The Scientific Imagination,* pp. 48–50. Millikan's recollection about the train trip home is in *Autobiography,* pp. 91–92.

149 "it has not yet been possible": *The Scientific Imagination,* p. 50.

149 "I saw a most beautiful sight": Harvey Fletcher, "My Work with Millikan on the Oil-Drop Experiment," *Physics Today* (June 1982): 45.

153 Stokes's law (named after the nineteenth-century scientist Sir George G. Stokes) describes how small spherical objects fall in a viscous medium like water or air. Millikan later adjusted the equation so that it applied more closely to objects as tiny as his oil drops.

154 "subelectrons": The physicist was Felix Ehrenhaft of the University of Vienna.

154 "I never would have believed it": Fletcher told the story in "My Work with Millikan," p. 46.

154 Millikan and Fletcher wrote up the results: "The Isolation of an Ion, a Precision Measurement of Its Charge, and the Correction of Stokes's Law," *Science* 30 (September 1910): 436–48.

154 "Very low something wrong": *The Scientific Imagination,* pp. 70–71, and "In Defense of Robert Andrews Millikan," *Engineering and Science* 63, no. 4 (2000): 34–35.

155 *"He who has seen that experiment": Autobiography,* pp. 96–98.

155 1.5924×10^{-19} coulombs: or 4.774×10^{-10} statcoulombs.

155 Fletcher told his story in "My Work with Millikan."

155 Millikan's patronizing manner: See, for example, *Autobiography,* p. 70. David Goodstein gives other examples in "In Defense of Robert Millikan."

156 The controversy over Millikan's data is described in the essays by Holton and Goodstein.

ACKNOWLEDGMENTS

I DON'T know how this book could have been written without so many good libraries around me. First is the beautiful Meem Library, just up the hill at St. John's College, designed by the southwestern architect John Gaw Meem and filled with classics in the history of science from Ptolemy's *Almagest* to Millikan's *The Electron.* I was actually able to find a facsimile edition of Albert Michelson's handwritten notes from his 1878 measurement of the speed of light. Just as special is the Santa Fe Public Library downtown—its reading room is another architectural treasure—where the reference librarians helped me secure several interlibrary loans. The farthest I had to stray from Santa Fe was to the University of New Mexico in Albuquerque, where the old bound journals are still on the open shelves and not relegated to the prison of microfilm.

Early on, the enthusiasm of St. John's president, John Balkcom, was an inspiration. I also thank Hans von Briesen, the school's former laboratory director, who gave me my first experience with the Thomson and Millikan experiments, and William Donahue, Peter Pesic, and Ned Walpin, faculty members who made insightful comments on the manuscript. I'm grateful to Owen Gingerich and Gerald Holton at Harvard and John Heilbron at Berkeley for their advice. Daniel Todes at Johns Hopkins made many helpful observations about Pavlov, as did Roald Hoffmann at Cornell about Lavoisier.

As always, thanks go to my friends who volunteered to be early readers: Patrick Coffey, Louisa Gilder, Bonnie Lee La Madeleine, David Padwa, and Ursula Pavlish. A microscopic read by Cormac McCarthy compelled me to expunge semicolons and underworked commas (a few of which have snuck back in). In the final stage, the book was greatly improved by the sharp scrutiny, erudition, and good sense of Mara Vatz and the artistry of Alison Kent.

This is the sixth book I've had the good fortune to do with Jon Segal at Knopf and the third with Will Sulkin at Jonathan Cape and Bodley Head. Their counsel and encouragement are invaluable, as are those of my agent, Esther Newberg, who has been there from the start. At Knopf I would also like to thank editorial assistant Kyle McCarthy, designer Virginia Tan, copy chief Lydia Buechler, and production editor Kathleen Fridella for their skill and patience in turning a manuscript into a book.

INDEX

Page numbers in *italics* refer to illustrations.

Index

ILLUSTRATION CREDITS

3 Galileo Galilei, by Ottavio Leoni. Wikimedia Commons.

7 An early nineteenth-century demonstration of the inclined plane experiment. Drawing by Alison Kent.

10 A page from Galileo's notebook. Folio 107v, vol. 72. Reproduced by kind permission of the Ministero per i Beni e le Attività Culturali, Italy/Biblioteca Nazionale Centrale. Firenze. This image cannot be reproduced in any form without the authorization of the Library, the owners of the copyright.

15 Galileo's finger. By permission of Istituto e Museo di Storia della Scienza.

17 William Harvey, by Willem van Bemmel. From a wood engraving by Jacob Houbraken of a painting. In Roswell Park, *An Epitome of the History of Medicine* (Philadelphia: The F. A. Davis Company, 1897), p. 156.

20 The Anatomy Theater of Fabricus. From a seventeenth-century engraving in Jacopo Filippo Tomasini, *Gymnasium Patavinum* (Udine: Nicolas Schiratt, 1654). Public domain.

24 Cross section of a human heart from *Gray's Anatomy*. Henry Gray, *Anatomy of the Human Body*, 20th edition (Philadelphia: Lea & Febiger, 1918).

29 Blood vessels, from Harvey's *Motion of the Heart*, 1628.

31 Isaac Newton, by Sir Godfrey Kneller, 1689.

33 *An Allegorical Monument to Sir Isaac Newton*, by Giovanni Battista Pittoni. By permission of the Fitzwilliam Museum, Cambridge, England. In 2005, when I went to see the painting, it had been moved to a stairway at the Trinity College Library.

35 Viewed under a microscope, "a small white spot of hairy mould." From Robert Hooke, *Micrographia*, 1665. Schem: XII, between pp. 124 and 125.

37 A lens sandwich used to show Newton's rings. Redrawn from a diagram in the Harvard Natural Sciences Lectures. Wikimedia Commons.

39 Newton's experiment with his own eye. A page from his notebooks. MS ADD 3975, p. 15. By permission of the Syndics of Cambridge University Library.

40 Newton's drawing of his *Experimentum Crucis*. From Newton's *Corre-*

spondence I, p. 107. (MS 361 vol. 2. fol, 45.) By permission of the Warden and Fellows, New College, Oxford.

45 Antoine-Laurent Lavoisier. Wikimedia Commons.

46 Incinerating diamonds. Bridgeman Art Library.

50 A pelican flask. John French, *The Art of Distillation* (London, 1651).

51 Marie Anne Pierrette Paulze. An engraving by Arents from a pastel portrait by an unknown artist. From Édouard Grimaux, *Lavoisier, 1743–1794, D'après Sa Correspondance, Ses Manuscrits, Ses Papiers De Famille Et D'autres Documents Inédits*, 3. éd (Paris: F. Alcan, 1899).

53 Burning litharge in a jar with a magnifying glass. From Lavoisier, *Mémoires De Chimie* (Paris, 1805).

56 Heating mercury in a "flamingo flask." Lavoisier, *Elements of Chemistry* (Paris, 1784). Plate 4, figure 2.

60 Luigi Galvani. Wikimedia Commons.

62 Symmer's socks. In Jean-Antoine Nollet, *Lettres sur l'électricité III* (1767).

63 An eighteenth-century static electricity machine. From Jean-Antoine Nollet, *Essai sur l'électricité des corps* (1750).

63 Benjamin Franklin's drawing of two Leyden jars. Benjamin Franklin, *Experiments and Observations on Electricity* (London: E. Cave, at St. John's Gate, 1751).

64 Muscular contractions caused by lightning. From Luigi Galvani, *De Viribus Electricitatis in Motu Musculari Commentarius*, 1791. Table II.

65 Static electricity and frogs' legs. From Luigi Galvani, *De Viribus Electricitatis in Motu Musculari Commentarius*. Table I.

71 Galvani's experiment without external conductors. From Marc Sirol, *Galvani Et Le Galvanisme. L'électricité Animale* (Paris: Vigot frères, 1939).

73 Volta's electrical pile. From Alessandro Volta, *On the Electricity Excited By the Mere Contact of Conducting Substances of Different Kinds. In a Letter From Alexander Volta to Sir Joseph Banks, Bart* (1800).

75 Michael Faraday. Project Gutenberg Archives.

77 Lady Ada Lovelace. Dated 1838. Wikimedia Commons.

78 Oersted's experiment. From Faraday's *Forces of Matter* (1868), p. 85.

79 From Faraday's diary, a wire rotating around a magnet. Sept. 3, 1821, p. 50.

80 Faraday's drawings of an induction ring. From his diary. Aug. 29, 1831, p. 367.

84 Polarization by reflection and through a polarizing crystal. Redrawn based on a diagram in *Scientific American*, July 1955.

A NOTE ABOUT THE AUTHOR

GEORGE JOHNSON writes about science for *The New York Times, Scientific American, Wired, Slate,* and other publications. His most recent books are *Miss Leavitt's Stars: The Untold Story of the Woman Who Discovered How to Measure the Universe* and *A Shortcut Through Time: The Path to the Quantum Computer.* Others include *Strange Beauty: Murray Gell-Mann and the Revolution in Twentieth-Century Physics* and *Fire in the Mind: Science, Faith, and the Search for Order,* which were finalists for the Royal Society Book Prize. A winner of the AAAS Science Journalism Award, he is codirector of the Santa Fe Science Writing Workshop and a former Alicia Patterson fellow. He lives in Santa Fe and can be found on the Web at www.talaya.net.